Plain Talk About Drinking Water

Plain Talk About Drinking Water

Questions and Answers About the Water You Drink

Fourth Edition

2001
American Water Works Association
Dedicated To Safe Drinking Water

American Water Works Association
6666 West Quincy Avenue
Denver, CO 80235-3098
1-800-926-7337

Acquistions Editor: Colin Murcray
Project Manager: Mindy Burke
Cover Design: Carol Magin Stearns

Library of Congress Cataloging-in-Publication Data
Symons, James M.
    Plain talk about drinking water : questions and answers
    about the water you drink.--4th ed.
        p. cm.
    Includes index.
    ISBN 1-58321-126-8
        1. Drinking water--Miscellanea. 2. Water quality--
Miscellanea. I. Title.

TD353 .S96 2001
363.6'1--dc21   2001053788

To my wife, Joan, and to our children,
Andy, Linda, and Julie, for their patience
throughout the years; and to my mother and
father for their lifelong support

Contents

Chapter 3   Home Facts—*continued*

Chapter 4   Conservation

Chapter 5   Sources

Chapter 6   Suppliers

Chapter 7   Distribution

Chapter 8   Regulations and Testing

Appendix A   Where Can I Get
More Information?

Appendix B   Information About Inorganic
Chemicals Found in Drinking Water

Preface

During my professional career, I have received many telephone calls from people concerned about their drinking water. In addition, friends, neighbors, and relatives have sought answers from me about the water they drink. This curiosity is understandable and proper, as drinking water is so vital to life. Yet to many people it is a mystery: Where does it come from? Is it safe? What about newspaper and magazine articles that describe the hazards of drinking water? Are advertised products for the home any good?

I prepared answers to the most common of these questions. Presenting the facts in nontechnical language for the average reader, making it easy to become an informed consumer. As an additional aid to the reader, the few scientific terms used are included in a phonetic guide in Appendix E. The questions are organized by categories; for example, all questions relating to the health (safety) aspects of drinking water are in one chapter of the book. The index makes finding the answers to your specific concerns still easier. In addition, five appendices—one listing sources of additional information, the second containing information about inorganic chemicals found in water, the third discussing interesting information about water in general, the fourth listing the questions themselves, the fifth a pronunciation guide—are included.

The list of questions provides an easy way to quickly get an overview of all the questions in a section or to locate a particular question. Some questions and answers are related; in these instances, cross-references are provided at the end of the answers to help you locate the related topics. And, the terms listed in the final appendix, which is the pronunciation guide, appear in bold text within the body of this book.

One note of caution. Although I have made every attempt to be accurate, these answers must be general in nature; thus, I cannot cover all variances caused by local conditions. Wherever possible, I suggest additional sources of information for these.

In addition, although most of the material is generic, some information is specific to conditions in the United States, Canada, or both. I appreciate the assistance provided by Ken Roberts, formerly of the Ontario Ministry of the Environment (currently with W20 in Mississauga, Ontario), and his staff who provided information on the Canadian situation in the answers to several of the questions.

No project like this can be completed alone. I would like to thank my friends and relatives who helped ensure that my answers were understandable to the general public: Marianne and Bill Myers, Lisa and Andy Symons, Linda Raab, Joan Symons, Nancy Logsdon, and Doug McIlroy.

I am also grateful to the professional colleagues who assisted me: Ray Taylor, Gary Logsdon, Tom Sorg, Mike Schock, Ed Geldreich, Irwin Kugelman, Jim Kreissl, Louis Simms, Warren Norris, Bob Hoehn, Jack Matson, Tom Pankratz, Roger Eichhorn, Dennis Clifford, Wendell LaFoe, Raanana Levin, Susan Gregory, Joe Harrison, Mark Dickson, Tom Love, Vanessa Leiby, Dick van der Koiij (who provided me the book *Laus Aquae*, from which the quotes that open each chapter were taken), Ed Ohanian, M. G. Christie, Mead Noss, Ric Jensen, Bill Lauer, George Symons, John Hoff, Roger Hulbert, Monica Baruth, Ron Packham, Katie McCain, Bill Knocke, and Ted Cleveland. In addition, I am indebted to the California Water Service Company for giving me permission to use some material from *Straight Talk on Water Quality*, a question-and- answer booklet for its customers.

I wish to thank my reviewers of the second edition of *Plain Talk About Drinking Water*—Jack W. Hoffbuhr, C. William Myers, Thomas J. Sorg, and Raymond H. Taylor—each of whom contributed much of their valuable time to assist me and to improve the final product.

Acknowledgments for the third edition include Karen Zicterman, Roger Hulbert, Don Reasoner, Joe Harrison, Stuart Long, Roger Eichhorn, Ray Taylor, Jack Cleasby, Mark Wiesner, Katie McCain, Kim Fox, Martha Panasar, Ron Packham, Linda Raab, Jon Stern, Frank Ferguson, Ruth Rosensweig, Heather Symons, Joan Symons, Larry Witte, Dick Bull, Doug McIlroy, John Wright, Vera Smart, Fred Pontius, and Chuck Simonds.

Acknowledgments for the fourth edition include Mike Schock, Stephanie Cimeno, Gary Burlingame, Stuart Krasner, Mel Suffet, Mike McGuire, Kim Fox, Bob Hoehn, Jim Mallory, Debi Huffman, Joe Harrison, Regu Regunathan, Tom Sorg, Joan Symons, and Nancy Zeilig.

Finally I acknowledge many current and former employees at the American Water Works Association who helped me, particularly Nancy Zeilig, Mary Kay Kozyra, Jon DeBoer, Mindy Burke, Norm Udevitz, Mark Scharfenaker, and Colin Murcray. This was a fun project for me; I hope it is informative and entertaining for you.

Jim Symons

How the Water Spirit Got Her Power

Once, long ago, Sun was the ruler of all the earth. Next to him, the other spirits were as the sparrow beside the grizzly bear. So the spirits had a secret meeting and elected the water-spirit to approach the Sun to give up some of his power.

Water went to Sun, and formed a clear, deep pool at his feet. When Sun saw his own face reflected in the pool, he was so delighted that he promised Water anything she wanted. When she demanded some of his power, he realized that he had been tricked, but according to his word, he gave power to all of the other spirits. Water, for her part, got more than anyone, and became, next to Sun, the most powerful force on earth.

—From a plaque next to Takakkaw Falls, Yoho National Park,
British Columbia, Canada

1
Health

Of this we may be sure: man must eat to live, and the problem of food will always be inextricably associated with water.

—Thompson King, *Water, Miracle of Nature*

GENERAL

1. Is my water safe to drink?

A definitive answer for countries as large as the United States and Canada is impossible, of course, but for the most part, yes. Nearly all public water supplies in the United States meet the US Environmental Protection Agency's standards for safe drinking water.

Standards are typically numerical limits on the concentrations, or amounts, of a particular contaminant. In cases where a contaminant cannot be readily measured, such as particular microbiological organisms that can sicken humans, water supplies must provide specific treatment, such as disinfection and filtration, to ensure safe water. *Note: Diesease producing microorganisms are often called germs; the scientific term is **pathogens**.*

Small water systems generally have more trouble meeting these standards than do larger cities. Smaller utilities frequently have a small rate base, so they can have difficulty raising money if corrections are needed. Such standards do not usually apply to private wells used by individual households.

Similar conditions exist in Canada.

(See Questions 25, 28, 187, 189, and 190 for related information.)

2. What is the definition of safe water? I've heard it called "potable water." How is "potable" pronounced?

Water is considered safe to drink if it meets or exceeds all of the federal, state, and provincial standards that are legally enforceable. As mentioned in Question 13, in the United States if your tap water does not meet any one of the standards, your water supplier must notify all its customers of the problem. Water is called **potable** when it is safe to drink. "Potable" rhymes with "floatable."

3.   How can I consider tap water safe after so many got sick in Milwaukee in 1993?

The waterborne-disease outbreak that occurred in Milwaukee, Wis., in spring 1993 made national news because of the number of people who got sick—about 400,000—and because of the number of deaths—about 100. The **organism** that caused the problem is called *Cryptosporidium,* an *intestinal protozoa*. One investigator likened the Milwaukee situation to a plane crash, an accident in an otherwise safe industry caused by a combination of unlikely events (see Question 4). Just as something can go wrong with an airplane, something went wrong in Milwaukee.

Lessons were learned from this tragedy, however, and many utilities in both the United States and Canada have improved their practices to reduce the chances of something like this happening again. Also, the US Environmental Protection Agency has since tightened its standards on surface water treatment to reduce the allowable level of the cloudiness of the treated water. Cloudiness of water is measured by a test for **turbidity,** and the water supplier has sensitive instruments that can detect slight changes in cloudiness, changes you could not detect by looking at the water. Measuring this cloudiness is important, because if it goes up a little, this indicates that treatment effectiveness is declining slightly. It is at these times that *Cryptosporidium* might slip through the plant and into the drinking water.

AWWA is recommending that drinking water treatment plants treating surface water should try to produce water that surpasses the federal standards, and that water suppliers should focus on maintaining 100 percent treatment reliability by having backup equipment, chemical feeders, and so forth. All these steps will minimize the possibility of another outbreak such as occurred in Milwaukee.

(See Questions 31–43, and 170 for related information.)

4. What happened in Milwaukee to make so many people sick?

No one knows for sure, but several things happened that could have caused the problem. The first event was a storm that washed contaminated runoff from several sources, including both agricultural areas and municipal sewers, into the source water for one of

Milwaukee's treatment plants. The agricultural areas included both pastureland and cattle-holding areas. Cattle wastes frequently contain the parasites that cause the disease **cryptosporidiosis**. The heavy rainfall also caused the sewers to overflow, putting municipal wastewater directly into the river used as a drinking water source. At this same time, the water treatment plant personnel were changing the chemicals they used to treat the water. During a short period, while they were determining the best dose of these chemicals, plant performance declined. Consequently, one probable cause of the outbreak was felt to have been *Cryptosporidium* slipping through the water treatment plant at that time.

5. From what I hear and read in the media, tap water certainly doesn't sound safe, yet water suppliers say it's okay. Who's right?

You may be surprised to learn that even though they seem to disagree, environmentalists, public health officials, the US Environmental Protection Agency (USEPA), Health Canada, provincial agencies, and your local water supplier all want the same thing: safe drinking water. For example, the water supply industry and USEPA have formed a voluntary Partnership for Safe Water that has as its goal maximum protection against the passage of **pathogens** through the water treatment plant.

The environmentalists are anxious to keep your attention on water quality problems so you will support your local water supplier's efforts to improve tap water quality. Your local water supplier wants your support, too, but often finds some of the environmentalists' reports in the media misleading. In one example related to chemical contamination, an August 1995 report by the Environmental Working Group said it had sampled some tap waters in the Midwest and that the levels of two weed killers used on corn and sorghum fields were above the federal government's allowable level during the growing season. That was not quite right. USEPA rules tell a water supplier to test one sample every three months and average these results together to get a pesticide level for the entire year. This average level over a year is the number that is compared to the government standard (which is based on being exposed to these chemicals for a lifetime, including a large margin of safety). In this case, the average levels of these chemicals were not above USEPA standards, so USEPA rules were not being violated, contrary to what was in the media. On the other hand, these chemicals *are undesirable* in drinking water, so suppliers urge restricted use on crops and some are considering treatment specifically designed to remove those chemicals.

Another source of confusion relates to the sampling rules. If a water supplier does not test the water as frequently as is called for in the rules, this is a violation. A sampling frequency violation does not necessarily mean that the tap water quality is poor; it **does**, however, mean that the supplier does not know enough about the quality of water as it should.

Media reports of violations must be studied carefully. If you're concerned about your local situation, your water supplier can tell you about its record of complying with USEPA rules, and if there are problems, what it is planning to do about them. Remember, you must be notified if any USEPA rules are violated.

(See Chapter 8 for related information.)

6. Is it true that tap water quality is getting worse?

It might seem that way from what you read and hear, as chemists and microbiologists are able to find more contaminants than ever before, but actually the opposite is true. Water suppliers must meet many more rules today than they did a few years ago, and standards for many of the regulated chemicals and microbes are more strict than they were a few years ago. Tap water quality is improving, although it is being talked about more because the general public is more aware of water quality issues and is demanding more information.

(See Questions 187 and 190 for related information.)

7. If most tap water is safe, why are engineers and scientists still doing so much research and why is the federal government thinking about more regulations?

As mentioned in Question 186, water that meets all of the federal, state, and provincial regulations is considered safe to drink, but it is not risk-free. Risk-free tap water would be too expensive, so in setting standards, the government chooses an acceptable risk (very small). Everyone wants to keep lowering this risk even further while not spending too much money. This is one of the goals of research and new government regulations. The other is to deal with any new potential problems that might be uncovered, such as the discovery of new contaminants.

8. Can I tell if my drinking water is okay by just looking at it, tasting it, or smelling it?

Unless germs occur along with other signs of contamination such as sewage smells, they generally cannot be detected by looking at, tasting, or smelling water. As noted in the answer to Question 51, some very good tasting stream or spring water can contain germs.

Chemicals are a somewhat different matter. Although chemicals cannot be seen in water, many do impart tastes or odors. Common examples of organic chemicals are dry cleaning solvents and the gasoline additive **methyl-tertiary-butyl-ether** (MTBE). Fortunately, the US Environmental Protection Agency's health limits and proposed limits on these chemicals are much higher than the amounts that cause tastes and odors. This means that even if you taste or smell an organic odor, the water can still be safe to drink. You should report any problem like this to your water supplier.

Other chemicals that can cause tastes or odors are discussed in the answer to Question 82. Your drinking water should be clear and nearly taste-free and odor-free. Report any changes to your water supplier immediately, as they could signal a problem.

(See Questions 51, 58, and 82 for related information.)

9. How do I find out if my water is safe to drink?

If you do not have your own private well, you can contact your water supplier and ask if the water meets federal, state, or provincial standards. If it does, the water is safe to drink. (This is a good practice when you move to a new location.) You can also ask your city, county, state, or provincial health department. In the United States, federal law states that consumers must be notified if violations of regulations occur. Federal law also requires most water utilities to provide customers with an annual water quality report called a Consumer Confidence Report. Canadian law requires water quality problems be reported to public health agencies, but does not require violations to be reported to consumers. If you have your own private well, you are responsible for having it tested yourself. Once it has been tested, you can discuss the results with your local health department.

10. I received a drinking water quality summary from my water supplier. What key information should I look for?

The formats vary from supplier to supplier, but most of these reports contain a table of constituents found in the local drinking water. For each constituent, the table usually shows the US Environmental Protection Agency's maximum contaminant level (MCL) and the amount found in your drinking water. If the amount in your supply is the same as or less than the MCL, your supply is all right. This will be the case in the majority of the situations. You will find other educational information in the report as well.

(See Question 188 for related information.)

11. My water supplier's annual drinking water quality summary report shows the amounts of some constituents as "not detected" (ND) or "below detection limits" (BDL). Is this the same as zero?

No. A level of zero would mean that not any of that constituent was in the water—not even one molecule, in the case of chemicals. Since the testing instruments cannot measure that small an amount, only the smallest amount of material that will cause a reading is known. If no reading is obtained, then any trace of that constituent in the sample was too small to register. The instrument operator then reports ND or BDL for that constituent. Because today's instruments are very sensitive, a report of ND or BDL means that very little of that constituent was in the water, but no one knows exactly how little.

12. My water supplier's annual drinking water quality summary report shows the maximum contaminant level (MCL) for some constituents to be "treatment technique," not a number. What does treatment technique mean?

In the Safe Drinking Water Act, the US Environmental Protection Agency (USEPA) is directed to control contaminants related to human health in drinking water by either setting a maximum contaminant level (the most common case) or, under certain circumstances, by prescribing one or more treatment methods that must be used by the water supplier. When a treatment method is specified by USEPA, it must be executed according to USEPA rules.

(See Questions 184 and 189 for related information.)

13. I've received a notice from my water utility telling me that something is wrong. What's that all about? What is a boil water order?

In the United States, if a supplier violates any federal drinking water standard, the utility, by law, must notify the customer. The idea behind this requirement is that the consumer has a right to know if the supplier is complying with all the standards. In addition, water systems are required to provide customers with annual water quality reports. These reports must describe the source and quality of the water it provides.

All violations are important, of course, but they are not all equally important. For instance, if the problem reported to customers is that the water supplier has not sampled the water as frequently as required by the regulations, this does not *necessarily* mean that the quality of the water is poor.

One violation involving insufficient sampling, improper reporting, or water quality does not mean the water is unsafe to drink. It does mean, however, that the water utility should improve its operations. On the other hand, if coliform bacteria (which indicate the possible presence of germs) are found in the drinking water, customers will generally be issued a *boil water order*—that is, they will be told to boil their water until further notice. In 1996, regulators and utilities agreed on uniform guidelines on when to issue a boil water order and when to stop it. Remember, if you have a boil water order in your area, be sure to throw all of your ice cubes away. They may have been made with contaminated water.

While consumers in Canada may not be directly notified of problems, provincial health or environmental agencies issue boil orders when warranted.

(See Questions 25, 29, and 184 for related information.)

14. I live in an apartment and don't get a water bill. How will I find out if there are any problems with my tap water?

Serious problems, called *acute violations*, must be broadcast on the radio and television and published in general circulation newspapers. Less serious problems must be published in general circulation newspapers and may also be otherwise posted or be sent to everyone who pays a water bill, in your case the apartment owner or manager.

Talking to your apartment manager and asking him or her to post anything included with the water bill is a good idea, however. Even when there are no problems, water suppliers often include valuable information (often called bill stuffers) with the water bill. Apartment dwellers should be able to see this information.

15. Beyond the Milwaukee disaster, does anyone actually get sick from drinking water?

Each year, several thousand cases of sickness in the United States and Canada are traced directly to drinking water. These illnesses (usually characterized by vomiting and diarrhea) are most often caused by inadequate disinfection of the water. Deaths are rare.

However, it is believed that many cases of waterborne illness go unreported. So, the true extent of the problem is unknown. Several studies are underway to better characterize the situation.

(See Questions 26 and 39 for related information.)

16. Is tap water safer in one area of a community as compared with another?

Rarely. All the tap water must meet all federal, state, or provincial requirements. In cities with a single source of water, everyone gets the same. Other communities have more than one source, so different parts of town get different quality water, but all the water must be safe to drink. The condition of the pipes and the flow patterns of water may be different in different areas, and this may cause some differences in water quality, although it usually does not affect water safety. To keep the distribution pipes clean, a water utility will flush them periodically. This practice may cause a temporary change in water quality. If you notice a change in water quality, you should notify your water supplier immediately.

(See Questions 173 and 174 for related information.)

17. I have read about animals dying after drinking reservoir water. If this can happen, how can I be sure my drinking water is safe?

Sheep, dogs, birds, or even cattle can die after drinking water heavily laden with microscopic organisms called **cyanobacteria**. These organisms are blue-green in color and often are mistakenly called blue-green algae. Heavy growths (called blooms) of these organisms sometimes occur in reservoirs during warm weather and can form a heavy scum on the surface. Some, but not all, cyanobacteria contain chemicals that are toxic to birds and animals. If the cyanobacteria population is large enough, the birds and animals that drink this water can die.

Water suppliers usually treat their reservoirs to prevent heavy growths of cyanobacteria, and any cells or toxins that reach the water treatment plant are readily removed there. Cyanobacteria also produce chemicals that cause tastes and odors. Whenever cyanobacteria grow in large numbers, the water will be treated more aggressively. This extra treatment will remove any cyanobacteria products of concern.

Water utilities are also studying cyanobacterial toxins—**microcystin** is the toxin of most concern—to improve their ability to prevent any problems. In 1998, the World Health Organization selected a limit of 1 mg/L (see Question 188 for definition) for microcystin. Microcystin is currently (2001) not regulated by the US Environmental Protection Agency, but is under study.

18. People are allowed to swim and go boating in our reservoir. Should I worry about this?

Although the swimmers and boats do add some pollution, when this pollution is diluted by all the water that is in the lake or reservoir, it usually doesn't amount to much. In addition, because the water is thoroughly treated before it comes to you, any contamination will have been removed. Fires, litter, and stormwater runoff can cause far more trouble than this kind of pollution.

19. Is tap water suitable for use in a home kidney dialysis machine?

No, not without further treatment. In a kidney dialysis machine, the water used is brought into close contact with the patient's blood. Thus, the quality requirements are far stricter than those for ordinary drinking water. Aluminum, fluoride, and **chloramine** are examples of substances that are okay if found in drinking water but are not acceptable in water used for kidney dialysis. Water suppliers cannot be expected to meet these strict requirements, so the water is further treated immediately prior to use in the dialysis machine. Kidney dialysis centers are kept informed about water quality by water suppliers and are able to give advice on this matter.

(See Questions 26, 67, and 73 for related information.)

20. Is it true that people who take antacids regularly are more likely to get sick from drinking water?

This has not been proven. Because the acid in the stomach that aids digestion tends to kill any germs a person might drink, some health professionals think that antacids, by destroying stomach acid, decrease this natural protection against germ disease. Presently, this is speculation, not fact.

21. Is the recommended six to eight glasses of water needed each day to maintain good health required to be tap water, or are other drinks okay?

Juice, milk, and soft drinks are almost all water, so they do count toward the required total daily fluid intake. Nutritionists often recommend tap water, however, because some other beverages contain chemicals like caffeine and alcohol that cause one to lose water. These are called **diuretics**, and thus, they do not help maintain fluid balance as well as other drinks. Tap water does not have these chemicals, so it is a safe recommendation, although other non-alcoholic drinks, including caffeine-free soft drinks, are fine. *NOTE: Decaffeinated coffee and tea do have some caffeine in them, so they are not as good as caffeine-free drinks.*

Older people sometimes do not drink enough liquids because their thirst mechanism is not strong enough. Thirst should not be an indicator of the daily need for liquids. Consumption of salty foods, diseases such as diabetes, and various medications all can affect a person's thirst sensation. Everyone needs fluids, whether they are thirsty or not. Finally, in proportion to body weight, babies need more fluids than adults. Consult with your doctor as to the water needs of your baby.

22. I'm moving or traveling to a higher altitude, and I'm concerned about dehydration. How much water should I drink?

Dehydration is a valid concern when adjusting to a higher altitude. Each person's reaction will be different, depending in part on how much of an altitude change is being made. The basic six to eight 8-ounce glasses of water should be the minimum, more if you are involved in strenuous activities. If you have any special health concerns, it is always a good idea to consult with your doctor before traveling to higher altitudes.

23. When I'm working in the yard, I'm tempted to take a drink from my garden hose. Is this safe?

No. A standard vinyl garden hose has substances in it to keep the hose flexible. These chemicals, which get into the water as it goes through the hose, are not good for you. They are not good for animals or pets either, so filling drinking containers for them out of a garden hose is not a good idea unless the water is allowed to run a while to flush the hose before using the water.

However, one type of hose on the market is made with a "food-grade" plastic that is approved by the US Food and Drug Administration and will not contaminate the water. Campers with recreational vehicles should use this type of hose when hooking up to a drinking water tap at a campsite. Check with a store that sells accessories for recreational vehicles.

Even a well-flushed vinyl hose or a food-grade plastic hose can cause problems, however. The outside thread opening at the end could be covered with chemicals or germs from a previous use.

24. Does the author drink tap water?

Yes, without reservation.

MICROBIAL CONTAMINANTS

General

25. Is my drinking water completely free of microbes?

No, but don't be alarmed; most microbes are harmless. For example, if you licked your finger, you would get microbes in your mouth, but you wouldn't get sick. Tap water usually contains harmless microbes. Microbes that can make you sick are called germs or **pathogens**. Tap water should be, and probably is, free of germs, however. Because most tap water is germ-free, pediatricians in metropolitan areas usually do not think it is necessary to boil tap water used in making baby formula. Check with your own pediatrician.

26. How are germs that can make me sick kept out of my drinking water?

A chemical called a *disinfectant* is added to drinking water at the treatment plant. Chlorine is the most common disinfectant used in the United States and Canada. Almost all the rest use a close relative of chlorine called chloramine. Chlorine and **chloramine** kill germs, but in this application do not harm humans. More recently, some water utilities have installed ozone as a disinfectant, and ultraviolet (UV) light has also emerged as a promising disinfectant. Your water supplier can tell you what is used in your water.

Although private well water is usually germ-free and not disinfected, it should always be tested at least once—annually is better— to uncover any possible germ contamination.

Some germs—for example, *Cryptosporidium*—are very difficult to kill using chlorine. This pathogen is rare in protected groundwater, and most water suppliers who use surface water, where this pathogen can be found, depend on filtration for removing them. Ozone and UV light, however, have been shown to inactivate this microorganism.

(See Questions 31–43, 69, 70, 145, 170, and 193 for related information.)

27. Viruses, bacteria, and protozoan cysts can all make me sick. Which is the hardest to kill?

The protozoan cysts are the hardest to kill, with *Cryptosporidium* being harder to kill than *Giardia*. *Cryptosporidium* is so hard to kill with chlorine that water suppliers that use surface water as their source depend primarily on filtration to remove this organism, instead of trying to kill it. As noted in the previous question, ozone and ultraviolet light have been shown to be effective at inactivating these germs. In general, viruses are more difficult to kill than bacteria, but the disinfection step in water treatment (see Question 26) is adequate to prevent either type of pathogen from reaching the consumer. For more information on *Cryptosporidium*, see the next section.

28. I'm told that I shouldn't drink my well water or that I need to boil it because my water has coliforms in it, but I'm also told that coliforms are harmless. Then I read that food poisoning can occur because of coliforms in meat. What are coliforms and what's going on?

Coliforms generally are harmless bacteria (a type of microbe) that live naturally in the intestines of humans and aid in the functioning of the body.

Water that contains coliforms is *not* safe to drink, however, not because of the coliforms, but because of the germs that possibly may be in well water when coliforms are found there. (Germs live in the body in the same place as the coliforms.) In fact, coliform organisms are called *indicator organisms* because their presence indicates the possibility of pathogen contamination.

NOTE: A few types of coliform bacteria do act as germs and cannot be ignored. These bacterial germs, called intestinal pathogenic **Escherichia coli** *0157:H7 by scientists (E. coli for short), have caused several food-borne and waterborne outbreaks of disease, with*

the loss of life primarily among senior citizens and
young children. Fortunately, such occurrences are rare.
(See Question 13 for related information.)

29. How can I kill all the germs in my drinking water?

Using a timer, bring the water to a full boil on a
stove or in a microwave oven, then boil it for one
minute. Because the boiling temperature of water
goes down about 2°F (1°C) for each 1,000 feet
(305 meters) you live above sea level, people living
at high altitudes should increase the boiling time.
For example, in Denver, Colo., which is more than
5,000 feet (1,525 meters) above sea level, boiling
time should be increased to three minutes.

Treating water in this way should be done only
in emergencies, because heating and boiling use a

lot of energy, create a
burn hazard for
children and the
elderly, and
concentrate some
chemicals (nitrates
and pesticides) if
they are in the

drinking water. However, the advantage of killing
the **pathogens** in an emergency outweighs the
slight disadvantage of concentrating the chemicals,
which results in only a minor worsening of water
quality.

Also, always be careful with boiling water! Let it cool in a safe place.

30. Could my drinking water transmit the AIDS virus?

There is absolutely no evidence that AIDS can be transmitted through drinking water. There is no danger from drinking water for three reasons. First and most important, you can't get AIDS by drinking the virus; it must get into the blood directly. Second, the virus is very weak outside the body and rapidly becomes noninfectious. Finally, even if present in water sources, the virus is easily killed during the disinfection step of drinking water treatment.

(See Question 26 for related information.)

Cryptosporidium

31. What are *Cryptosporidium* and cryptosporidiosis?

Cryptosporidium is a protozoan parasite that can live in the intestines of humans and animals (hosts). Outside of the hosts the microbe is protected by a shell called an **oocyst**, so it is like a seed of a plant, very tough and long lasting. Once swallowed, the microbe emerges from its shell and infects the lining of the intestine. When this happens, some people get a disease called **cryptosporidiosis**. In people, the usual time between swallowing this microbe and getting sick is from 2 to 10 days. The major symptom is severe watery diarrhea with cramping abdominal pain, which lasts about 10 days to 2 weeks for people with normal immune systems. Other symptoms can be nausea, vomiting, fever, headache, and loss of appetite. Thus, although cryptosporidiosis is an unpleasant disease, it is not a dangerous one to people with normal immune systems. *NOTE: Not everyone who gets infected gets sick. Sometimes people who have been infected with* Cryptosporidium *don't know it.*

32. Where do the microbes that cause cryptosporidiosis come from?

They are in the stools of infected animals such as cattle, lambs, wild animals, and people. The wastes from these animal and human sources get into the environment and then into surface waters, rivers, lakes, and streams from runoff over land (rain or melting snow and ice) and wastewater. *NOTE: Wastewater treatment does not remove these microbes. Thus, a water supplier that uses surface water for its source may find these disease-producing microbes (germs) in that source water.*

(See Question 156 for related information.)

33. What is the medical treatment for cryptosporidiosis?

Presently, no effective cure is available for cryptosporidiosis. People who have normal immune systems improve without taking antibiotics or other medicine. The treatment for the diarrhea is to drink plenty of fluids and to get extra rest. Doctors may prescribe medication to slow the diarrhea during the 10 days to 2 weeks usually required for recovery.

34. What should I do if I think I have cryptosporidiosis?

See your doctor. Because routine stool examinations frequently do not test for the protozoa that cause cryptosporidiosis, the laboratory needs to use special tests available just for this organism. It is important for those with impaired immune systems  to get medical attention as soon as they get sick.

35. Are all water systems affected by the threat from *Cryptosporidium?*

No. Protected groundwaters that are not mixed with surface water are usually free from these organisms. Thus, a water supplier that uses protected groundwater entirely should not have problems.

36. Is drinking water the only source of *Cryptosporidium?*

No. There are many other sources. For example, foods such as unwashed fruits and vegetables, especially from foreign countries, swimming pools, recreational water, day-care centers, and nursing homes are common sources. Studies on people's blood have shown that about one-third of the population has been recently exposed, indicating that there are many possible sources of exposure. Remember, for all of these sources, the common factor is contamination from stools of infected humans or animals.

37. My water supplier found some of the microbes that cause cryptosporidiosis in our water. Will I get sick if I drink water from the tap?

If you have a normal immune system, probably not. There are at least three reasons: First, using the current test for the microbes, the analyst can't tell whether they're dead or alive, so the microbes found might be harmless. Second, lots of water is sampled to look for the microbes, much more than would be in a glass of water. Finding a few in that large volume of water doesn't mean every glass of water has germs in it. Finally, although all people are presumed to be able to become "infected" from these germs, some people have such strong immune systems that they will not get the disease. For example, estimates have shown that if 1 million people each swallowed one germ, only 5,000 would get "infected," and only about two-thirds of these 5,000 would actually have the symptoms of the illness—a very small risk. Of course, if you are immunocompromised (a group with a high risk of infection), you should consider the precautions discussed in the answer to Question 41.

38. I'm worried about *Cryptosporidium* in my water. Should I have my water tested?

You can discuss this with your water utility. Costs for the test range from $400 to $500. Improved cartridge filters have increased the test's ability to recover **Cryptosporidium oocysts** from both surface water and drinking water, so the test for *Cryptosporidium* is better than it was. However, a single sample analyzed for the presence of *Cryptosporidium* may not give you a true picture of the quality of your water. Currently (2001), testing for common *indicator organisms* that point to the possible presence of germs, as well as alternative indicator organisms such as **enterococci** or male specific **bacteriophage** may give you more information about any potential contamination of your drinking water than a test for *Cryptosporidium* would.

(See Question 39 for related information).

39. Is my water supplier doing anything to prevent me from getting cryptosporidiosis?

If your supplier uses surface water as a source, probably yes, but you can call them for their specific plans. Many water suppliers that use surface water as a source are taking two actions: (1) they are trying to clean up their source water by controlling animal populations and human waste discharge, and (2) they are making their filters perform better. Filter performance is tested by measuring the cloudiness of the water coming out of the filter. The clearer the water, the better the filters are working. The US Environmental Protection Agency (USEPA) has recently tightened the federal standards for turbidity.

Many water suppliers in both the United States and Canada were trying to surpass the federal standard even before the Partnership for Safe Water was formed in the United States. The Partnership for Safe Water is a cooperative effort between the USEPA and the drinking water profession to improve tap water quality without new regulations. As mentioned in the answer to Question 5, the Partnership is focusing even more attention on the production of tap water free of the microbes that cause **cryptosporidiosis** and related diseases. Furthermore, the USEPA has recently established rules that require certain levels of removal of *Cryptospordium* through filtration. This should help keep the microbe that causes cryptosporidiosis from reaching your tap. In addition, public water systems in the United States have joined together and spent more than $10 million on research on how to control this organism, and this research continues.

40. How can I keep from getting cryptosporidiosis?

Avoid water or food that may be contaminated. Wash your hands after using the toilet and before handling food. If you work in a day-care center where you change or handle dirty diapers, be sure to wash your hands thoroughly with plenty of soap and warm water after every diaper change, even if you wear gloves. During community-wide outbreaks caused by contaminated water (a boil water order may be issued), boil all drinking water for one minute to kill the germs, three minutes if you live at a high altitude.

Avoid drinking water directly from lakes and rivers. Avoid unpasteurized dairy products. Avoid exposure to calves and lambs and places where these animals are raised. Wash your hands after contact with pets and after gardening and other contact with soil. If you are a caregiver for cryptosporidiosis patients, wash your hands after bathing patients, emptying bedpans, changing soiled linen, or coming into contact with stool material. If you have cryptosporidiosis, wash your hands often to prevent spreading the disease to other members of your household.

(See Question 13 for related information.)

41. I am immunocompromised. What should I drink?

If you are infected with HIV or have AIDS, are a cancer patient, are taking immunosuppressive drugs after a transplant, or were born with a weakened immune system, you need to discuss this matter with your doctor and consider the following. On June 16, 1995, the US Environmental Protection Agency (USEPA) and the Centers for Disease Control and Prevention (CDC) released guidance for people with severely weakened immune systems. Their statement pointed out that the risk of a person with a severely weakened immune system getting **cryptosporidiosis** from drinking water in the absence of an outbreak of the disease varies from city to city, depending on the quality of the city's water source (surface water is much more likely to contain the disease-producing microbe than groundwater) and the quality of water treatment. USEPA and CDC did recommend, however, that severely immunocompromised people who wish to take *extra* measures to avoid waterborne cryptosporidiosis can heat their drinking water to a rolling boil for one minute (three minutes at higher altitudes).

To be effective, boiling must be done all of the time and must include all water for drinking, cooking, brushing teeth, washing produce, and so forth. In short, any water you might swallow should be boiled. For more information, call the CDC's National AIDS 24-hr. Hotline at (800) 342-2437, or call the CDC's National Response Center Hotline at (800) 424-8802, and you will be referred to USEPA's Emergency Response Team.

42. Should I drink bottled water to prevent coming down with cryptosporidiosis?

Not necessarily. First and most important, your tap water probably does not have the germs that cause this disease; most tap water doesn't. Second, if you have a normal immune system, the disease is not life-threatening. Finally, bottled water suppliers typically do not test for the microbes that cause cryptosporidiosis and these germs can live

for weeks in water, even refrigerated water. Bottled water from a protected groundwater source, such as a well, would have less chance of becoming contaminated than would surface water or spring water that flowed over the land. Bottled waters with labels that mention treatment with ozone, filtration through an absolute 3.0 micron filter ("absolute" means no openings in the filter are bigger than 3.0 micron), or reverse osmosis are probably safe. For more information on bottled water, call the International Bottled Water Association Hotline at (800) 928-3711.

43. Will a home water treatment device protect me from the microbes that cause cryptosporidiosis?

Some will, and some won't. Distillation units, reverse **osmosis** systems, and filters with an "absolute rating of 3.0 microns" (sometimes called 3.0 micrometers), or those labeled as certified by reputable independent testing organizations following the American National Standards Institute (ANSI)/NSF Standard 53 Protocol for Cyst Removal, provide the greatest assurance of removing the germs that cause **cryptosporidiosis**. *CAUTION: The "nominal" three micron (micrometer) rating is not standardized, and many filters in this category may not remove the germs that cause cryptosporidiosis. An "absolute 3.0 micron" (micrometer) rating means that none of the openings in the filters is bigger than 3.0 microns, in contrast to a "nominal three micron" (micrometer) rating, which means that most of the openings are 3.0 microns (micrometers) in size; but some are smaller, and some are **bigger**. These bigger openings may let some of the germs through.*

Neither the US Environmental Protection Agency, the Centers for Disease Control and Protection, nor AWWA maintains a list of home treatment devices that are satisfactory. The Water Quality Association (WQA), Underwriters Laboratories, Inc., and NSF International can supply a list of filters that they have tested and that meet ANSI/NSF Standard 53. Check the filter label or call (800) 749-0234 (WQA), (800) 673-8010 (NSF International), or (847) 272-8800 (Underwriters Laboratories, Inc.) for substantiation of testing by a reputable independent testing organization. Of course, one minute of boiling (three minutes at higher altitudes) will always work.

Travel

44. If I travel overseas, in which countries is the water safe to drink?

Besides the United States and Canada, the water generally is safe to drink in western Europe, Australia, New Zealand, and Japan. In other countries, you should insist on carbonated bottled water for drinking and brushing your teeth. Some hotels fill containers with tap water, so be sure you get *carbonated* water so you can be sure the bottle was not filled from a local tap. Watch out for ice cubes and foods like gelatin that usually are made with tap water. Also, beware of salads; the greens are often washed in local tap water.

A *USA Today* article citing information from the Water Quality and Health Council of the Chlorine Chemistry Council reported that more than one billion people, one in every five on Earth, do not have access to safe drinking water. The following table shows the percentage of the population with access to safe drinking water in five sample countries:

Country	Percentage
United States	99
Mexico	72
Pakistan	56
Sudan	45
Ethiopia	18

The organization Water For People has a mission to provide sustainable drinking water for impoverished and developing countries. All donations are tax deductible in the United States and Canada. To help, contact them at:

Water For People
6666 West Quincy Avenue
Denver, CO 80235-3098
(303) 734-3490
http://water4people.org

Canadian residents may send their donations to:
    Water For People–Canada
    45 23rd Street
    Toronto, Ontario M8V 3M6
(See Question 1 for related information.)

45. When I'm traveling outside of areas where tap water is safe, getting bottled water is inconvenient and I worry about its quality. Is there something I can take with me to purify water?

There are portable mechanical water purifiers that are claimed to produce germ-free water, one glass at a time. These devices, which frequently combine a filter and a chemical disinfectant, should be registered as a "purifier" with the US Environmental Protection Agency (USEPA). To be registered, the device must be tested by USEPA according to methods contained in ANSI/NSF Standard 55, and its performance must be reviewed by USEPA. Any device that uses ultraviolet light or other methods that do not involve chemicals is not required to be registered by the USEPA but may have its performance checked by an independent organization following the methods contained in ANSI/NSF Standard 55.

Because a traveler cannot determine the proper conditions required to make sure that ultraviolet light will kill *Cryptosporidium* and *Giardia,* any device that uses *only* ultraviolet light should be used with caution.

46. How is water quality protected on airplanes?

There are three parts to airplane sanitation: (1) the quality of the water supplying the plane (called an *interstate carrier water supply*); (2) the protection of the water as it is put on the plane (called the *watering point*); and (3) the quality of the water as it is used on the plane (called an *interstate carrier*). The water quality from the interstate carrier water supply (where the water is collected) must meet all federal, state, and provincial standards. Watering point operators must make sure they do not spoil the water quality while they are putting it on the airplane. The US Food and Drug Administration is responsible for checking on the watering point operators. Once on the airplane, the airline must test for microbe contamination four times a year or it can provide the US Environmental Protection Agency with an operation and maintenance plan and request that it be substituted for the microbe testing.

47. How is water quality protected on cruise ships?

Most cruise ships operating out of US ports participate in the voluntary Vessel Sanitation Program run by the Centers for Disease Control and Protection (CDC). CDC personnel conduct unannounced inspections of the cruise ships twice

a year and rate the cleanliness of food and water on a 100-point scale. Drinking water items make up 27 of the 100 points. Most ships take water from the shore, disinfect it, and put it in tanks on

the ship (this is called *bunkering*). Some ships do, however, make drinking water from sea water. Whatever the source, the water then must be disinfected again before going to the passengers, and records must be kept to show that the disinfection was up to standards.

For more information about a specific ship call the CDC fax-back system at (888) 232-6789. You will receive a faxed summary of all of the recent inspections, giving the ship name, the date of the most recent inspection, and the score. If you have an Internet connection, access the address

<http://www2.cdc.gov>. Another method is to
write to Chief, Vessel Sanitation Program, National
Center for Environmental Health, 4770 Buford
Hwy., N.E. Mailstop F-16, Atlanta, GA 30341-3724
and ask for "The Green Sheet." Your travel agent
may also have a copy of "The Green Sheet." If you
want more details than are contained in "The
Green Sheet" about a specific ship, you can request
a copy of the most recent sanitation inspection
report from the Georgia address.

48. When I travel to a different place in this country, sometimes I have an upset stomach for a couple of days. Is this because something is wrong with the water?

This probably does not result from a problem
with germs in the drinking water. However,
waters with a high mineral content, particularly
sulfate, sometimes have a temporary laxative
effect if your body is not accustomed to them.
Therefore, the change in mineral content from
place to place sometimes does bother travelers for
a short time, until the body readjusts.

49. I've heard that households in the United States use a lot of water compared with other countries. Is this true?

Yes, U.S. residents use more water than residents of other countries, even countries that are equally well developed. In the United States, significant amounts of water are used for lawn and garden irrigation, automobile washing, and kitchen and laundry appliances, such as garbage disposals, clothes washers, and automatic dishwashers. The table below summarizes residential water use in selected countries.

Residential Water Use In Selected Countries

Country	Gallons of Water Used by One Person in a Year
United States	52,500
Canada	40,300
Malta	16,000
Poland	15,680
Belgium	13,260
Nicaragua	12,960
China	7,320
India	3,900

50. What water does the president of the United States drink at home and when traveling abroad?

Wherever he is, the president drinks commercial bottled water, not because of concern about the quality of tap water in the United States, but to avoid any temporary mild discomfort that may occur from variations in the quality of perfectly safe drinking water from place to place. Thus, to ensure a constant drinking water quality both here and abroad, bottled water is supplied to the president.

51. Is it okay for campers, hikers, and backpackers to drink water from remote streams?

Although chemical pollution is not normally a concern in remote streams because a person generally would not drink enough water to cause sickness, the answer is no.

These streams often contain *Giardia*, a pathogenic protozoan, and other protozoa such as *Cryptosporidium*, which can cause disease. These two germs cause illnesses called **giardiasis** or **cryptosporidiosis** that are characterized by severe diarrhea, which can last several weeks, sometimes even longer.

In addition, disease-causing bacteria from wildlife might also be present in remote streams.

(See Questions 31–34 for related information.)

52. What can campers, hikers, and backpackers do to treat stream water to make it safe to drink?

Any water that looks good enough to drink can be made microbiologically safe by boiling. One minute of *vigorous* boiling at sea level or three minutes at high elevations will kill all germs, disease-producing bacteria, viruses, protozoan cysts, and so forth.

Water disinfection tablets, available at drugstores and camping supply outlets, can be put into a glass of clear water. They take about five minutes to dissolve and release the disinfectant (see Question 26). These tablets *will not* protect against **Giardia** and **Cryptosporidium** (the primary concerns in most areas) or parasitic worms, however, and will not work well in cloudy or colored water. Thus, their use alone is discouraged.

Some portable water filters available on the market can provide effective treatment. One unit, which resembles a large-diameter soda straw, that contains a disinfectant and a filter, as well as other materials, has not been shown to be very effective, and the small-diameter copies of it work very poorly. Mouth suction is used to draw clear water through the unit. Even filters that will remove *Cryptosporidium* (rated "absolute" 3.0 microns— meaning that nothing larger than 3.0 microns in size will get through the openings) do not remove smaller disease-causing bacteria or viruses. Thus, disinfection tablets should be added to the filtered water before use for drinking. If the water is clear to start with, the mechanical purifiers described in the answer to Question 45 may work; just make sure they meet the requirements of the US Environmental Protection Agency. The only sure protection is to bring your own water or boil the "natural" water.

53. If I had no chemicals, no fire, and no filters, is there any way I could make stream or lake water safe to drink?

The World Health Organization (WHO) offers a suggestion. If you are in a warm climate, have plenty of sunshine, a black surface, and a clear plastic soft-drink bottle, fill the bottle about half full and lay it on its side on a black surface in the bright sunshine for about five hours. According to WHO, this will kill most, but not all, germs in clear water. The hotter the air temperature, the better. Although not perfect, this method, called "solar water disinfection," would be better than drinking untreated water.

CHEMICALS

General

54. Are all chemicals in my drinking water bad for me?

No. Some chemicals, fluoride at proper levels for example, are good for you, and some minerals are accepted by most to be beneficial in drinking water. In addition, many chemicals have no bad effect on your health.

Chemicals are not bad just because they are chemicals. For example, water itself is a chemical, and we depend on chemicals in food to keep us alive.

(See Question 67 for related information.)

55. Does drinking water contain calories, fat, sugar, caffeine, or cholesterol?

No.

56. Does drinking ice water burn fat?

A little. For example, an 8-ounce glass contains about 240 grams of water. Water with ice floating in it is at a temperature of 32°F (0°C). When consumed, the water warms to body temperature in the stomach, about 98.6°F (37°C). It requires about 8,800 gram-calories (240 × 37) to warm the water. Food calories are actually kilogram-calories, so drinking the glass of ice water uses 8.8 kilogram-calories. Thus, very little fat is burned in food energy terms.

57. Are chemicals that are found in drinking water naturally (rather than because of pollution) nontoxic?

Not necessarily. Many chemicals that occur in nature can be harmful to your health, and they can be present in water. A few examples are arsenic, radium, radon, and selenium. Also, some nontoxic natural chemicals combine with other chemicals to produce harmful chemicals (*reaction products*). Therefore, some "natural" chemicals must be watched closely by your water supplier. Water suppliers must test for dozens of chemicals regulated by the US Environmental Protection Agency. Some of these are "natural;" others are of human origin.

(See Questions 70, 75, 82, 187, and 188 for related information.)

58. I read that organic chemicals are bad for my health. What are they, why are they dangerous, and why doesn't my water utility just remove them?

Organic chemicals have mostly carbon atoms connected to hydrogen atoms (the element, not the gas). A common organic chemical in the home is sugar, so not all organic chemicals are dangerous. Food contains many beneficial organic chemicals essential to our life. *NOTE: You can't tell if something is an organic chemical just by looking at it.* For example, table salt looks a lot like sugar but does not contain carbon and hydrogen and so is an *inorganic* chemical.

Dangerous organic chemicals are found in products such as gasoline, cleaning fluid, pesticides, paint thinners, and car radiator fluid. They are dangerous if they get into your drinking water because many are cancer-causing chemicals—called **carcinogens**. The water treatment methods used by most utilities are designed to make drinking water clear and to kill germs. These treatment systems can't remove organic chemicals, which require different treatment methods. The US Environmental Protection Agency regulates many of these chemicals, and when these chemicals are found at levels that exceed standards, water systems install specific treatment to remove these chemicals.

(See Question 170 for related information.)

59. I'm told that our drinking water contains chemicals like cleaning fluid and benzene. What can I do about this while the water company improves treatment?

These chemicals are called *solvents*. The best thing to do is boil your water for 10 minutes (use a timer), either on the stove or in a microwave oven. About 20 percent of the volume will be lost during boiling, and this may concentrate other pollutants like nitrates and pesticides. Boil the water in a well-ventilated area. Mixing with an electric mixer or a blender for 10 minutes is also very effective, since vigorous mixing causes some of the chemicals to evaporate. Storing water in an open pan for 48 hours will also work. Aeration with an aquarium-type aerator or pouring the water back and forth between two containers are other, less effective treatments. Using an aerated faucet is ineffective. A more expensive option is to use bottled water as a temporary remedy, or you can install a home treatment device that has been tested and certified by an independent organization for "volatile organic chemical" reduction according to the methods described in the ANSI/NSF Standard 53.

These methods need to be considered only if your water company notifies you that the levels of solvents in your water exceed the drinking water standards.

(See Questions 110 and 111 for related information).

60. I've heard that nitrates are bad for infants and pesticides are bad for everyone. How do nitrates and pesticides get into my drinking water?

The US Environmental Protection Agency sets a limit for nitrates because high levels are associated with a rare blood condition in infants—commonly called the blue-baby syndrome because the baby's skin turns bluish after the baby drinks water containing nitrates. Nitrates have been a problem in some private wells. Nitrates and pesticides come from fertilizers and pesticides used on farms and on home gardens and lawns. Septic tank drain fields and wastes from animal feedlots are other important sources of nitrates. Rainwater takes these chemicals with it and contaminates groundwater and waterways.

As rain moves through the soil, microbes in the soil change the ammonia in the fertilizer and drainage from septic tanks and feedlots into nitrates.

If your water supply comes from a private well, you may want to use the services of a local water treatment dealer who can have your water tested and install proper equipment, if necessary, to remove nitrates and pesticides. The Water Quality Association has currently (2001) educated, tested, and certified more than 2,100 certified water specialists nationwide. Call (800) 749-0234 for information.

61. There is a blue-green stain where my water drips into my sink. What causes this?

This stain comes from the chemical copper. The copper probably is present in your home plumbing and is being dissolved into the drinking water. The conditions that cause copper in the water also can introduce lead into drinking water, and high amounts of either lead or copper can

cause health problems. If you have blue-green stains in your sink, you should call your local water supplier to discuss this. If your water is from a private well, have your tap water tested for lead and copper.

To clean the sink, check with your local hardware store for stain-removal products.

(See Question 63 for related information.)

62. Do hazardous wastes contaminate drinking water?

Yes, they may. As rainwater seeps down through a hazardous waste dump, it carries the chemicals with it to the groundwater. Some chemicals stick to the dirt particles and don't reach the groundwater very quickly. Other chemicals, such as cleaning fluid or gasoline, move down through the ground rapidly. Rain can also wash contaminants from a hazardous waste dump into surface waters, which can then seep into the groundwater and pollute it. This is one reason there are such strict rules on covers and liners in the areas where hazardous wastes are stored and abandoned locations are being cleaned up.

Leaking underground gasoline tanks at gas stations and the improper disposal of chemicals (for example, dumping old radiator fluid, metal degreasers, paint thinners, or paintbrush cleaners in the backyard) also may contaminate groundwater.

Surface waters in some regions are also contaminated by chemicals used for deicing roads in the winter. When it rains or the snow and ice melt, these chemicals wash into rivers, lakes, and reservoirs. To prevent this problem, some states post signs on roads that cross watersheds to inform the highway crews not to spread deicing chemicals in these areas. Improperly treated wastes from industrial plants may also pollute surface waters. Many water systems work hard to prevent such contamination, but if it occurs, these systems must treat their water to remove the chemicals.

(See Questions 156 and 161 for related information.)

Lead

63. How does lead get into my drinking water?

Not all drinking water contains lead. Where lead is present in pipes and in soldered connections, the lead may dissolve into the water while the water is not moving, generally overnight

or at other times when the water supply is not used for several hours. Faucets with brass or bronze internal parts may also be a major source of lead under these conditions.

The first water that comes from the faucet after long periods of nonuse may have lead in it. Hard water (see Question 101) sometimes picks up less lead than naturally soft low mineral content water because any water with low amounts of dissolved minerals tends to dissolve metal pipes, and hard water has a tendency to lay down a scale layer on the inside of pipes.

In the United States, lead is now banned in pipes and in solder, and the voluntary standard for lead in faucets has caused the plumbing industry to change manufacturing methods to reduce lead levels. An NSF mark on a faucet means that NSF International has tested the faucet and it meets the new standard. However, the piping systems in many cities and faucets still contain lead. Also, the ban on lead-based solder is inconsistently enforced. Consumers should insist that no plumbing repairs be done with lead-based solder and that it not be used in new homes.

In 1991, the US Environmental Protection Agency (USEPA) issued a new regulation for

controlling lead in drinking water. If high amounts
are found at the tap, corrective action by the
supplier is required and public education
materials accepted by USEPA must be delivered to
customers. In addition, the water supplier must
offer to sample the tap water of any customer who
requests it. The water supplier, however, is not
required to pay for collecting or analyzing the
sample, nor is the system required to collect and
analyze the sample itself.

Canada banned the use of lead pipe and lead-
based solder in 1990 and currently has a guideline
of 0.01 milligrams of lead per liter (abbreviated
mg/L—see Question 188 for definition) of water in
a sample collected from a flowing tap.

(See Questions 101 and 102 for related
information.)

64. How can I get lead out of my drinking water?

Not all homes have a lead problem, but if
testing has indicated a problem, if you think your
water is corrosive, or if you have rusty water or
blue-green stains in your sink, take the following
precautions.

Whenever water has not been used for a long
period of time—overnight or during the day if no
one is home—let the cold water run from the
faucet for two minutes (this is a long time) before

using any water for drinking or cooking (see NOTE below). Saving this water for other purposes such as plant watering is a good conservation measure. Letting the water run for two minutes will not flush out all the lead that got into the water while it was sitting in your plumbing, but it will improve the situation greatly.

Some home treatment equipment (lead-removing filters, reverse **osmosis** systems—sometimes called RO systems, and distillation units) removes lead dissolved in water. Check to see whether the performance of these products has been tested for lead reduction by independent testing and certifying organizations following the methods contained in the appropriate ANSI/NSF "Drinking Water Treatment Unit" standard.

You can get information on lead hazards by calling the National Lead Information Center (see Appendix A).

NOTE: It is difficult to know how long it takes the "fresh" water from the street pipes to arrive at the faucet. The time needed varies depending on your specific location, water pressure, whether you live in a single-family home or an apartment, and so forth.

If the water from your cold water faucet gets colder after it has run for a while, always leave the tap open until you feel the colder water. Otherwise, two minutes is usually enough for most homes. Flushing a toilet doesn't work; you must run the faucet you are actually going to use for drinking or cooking.

(See Questions 61, 86, and 96 for related information.)

65. How can I find out if my water is supplied through lead pipes?

The water main in the road is not usually made of lead—usually it's cement-lined cast iron or sometimes plastic—but the connection from the pipe in the street to the pipe to your house (often called a gooseneck because of its shape), the pipe connecting it to your house, or the pipes within your house might be. Contact your water supplier to find out what materials have been used.

In older homes, the household pipes might be lead, copper, or galvanized iron. Joints on lead pipes are usually very bulky compared with the relatively neat fittings of copper and galvanized iron. You can tell copper by the color (like a penny), but you might have to scrape off some paint to see the actual pipe material. Copper and galvanized iron give a more metallic sound when gently tapped with a small hammer. If you are in doubt, consult a reputable plumber.

Lead pipes are unlikely to be found in newer housing, as their use has been banned since 1986. Plastic is quite common now.

66. Is it safe to drink water from a drinking fountain?

Yes, usually. Some older, floor-standing water coolers contain lead-lined storage tanks, and there is a possibility that high amounts of lead could get  into the water in these tanks. The US Congress passed laws banning the use of lead in piping and in solder in 1986 and in storage tanks in 1988, and these regulations have improved the situation.

One section of the 1988 act encourages schools and day-care centers to test their fountains for lead. Canada banned the use of lead in pipes and of lead solder in 1990.

Letting the water run for a while before drinking from a drinking fountain minimizes the risk. However, because the storage tanks are about 1 quart (about 1 liter) in size, complete flushing may take a while.

Fluoride

67. Is the fluoride in my drinking water safe?

Yes. When added or naturally present in the correct amounts, fluoride in drinking water has greatly improved the dental health of American and Canadian consumers. Early studies suggesting that fluoride was a possible cancer-causing chemical proved to be incorrect. A 1993 report by the National Research Council of the National Academy of Sciences, *Health Risk of Ingested Fluoride*, states, "Currently allowed fluoride levels in drinking water do not pose a risk of health problems such as cancer, kidney failure, or bone disease." Excess fluoride in water is removed by the water supplier using special treatment.

For one reason or another, about 40 percent of Americans do not have adequately fluoridated water supplies, although fluoridation is mandatory in many states. Recently, the American Dental Association changed its recommendation on dietary supplements for children, advising no fluoride supplement be given to infants younger than six months old. You may want to talk to your doctor about this.

When present even in correct amounts, fluoride and the disinfectant **chloramine** do make water unsuitable for use in kidney dialysis machines.

Dialysis patients should check with their water supplier or dialysis center about their water source.

(See Questions 19 and 54 for related information.)

68. Will I lose the benefits of fluoride in my drinking water if I install a home treatment device or drink bottled water?

Certain types of home treatment devices will remove 85 to more than 95 percent of all the minerals in water, including fluoride. These are reverse **osmosis**, distillation units, and deionization units (not water softeners—they leave fluoride in the water). If you use one of these types of devices, consult with your dentist about fluoride and possibly your doctor about iodine supplements.

The situation with bottled water is less clear. One recent study showed many bottled waters contained very little fluoride, although a few contained adequate amounts. Unfortunately, even in those products, the fluoride level dropped to about 25 to 50 percent of the original value over a two-year sampling period, without a change in product name or label. If you are drinking bottled water, most likely you are not getting much fluoride. Remember, if you are using bottled water to make formula for your baby, be sure to talk to your doctor about using fluoride supplements.

(See Questions 96, 102, and 110 for related information.)

Chlorine

69. Is water with chlorine in it safe to drink?

Yes. Many tests have shown that the amount of chlorine found in treated water is safe to drink, although some people object to the taste. The US Environmental Protection Agency recently established maximum allowable levels of "residual" disinfectants, which are added to water as it enters the distribution system to protect against germs.

(See Questions 26, 82, 83, and 96 for related information.)

70. What is the link between chlorine and cancer?

Chlorine is added to drinking water to kill **pathogens**. While chlorine does provide this protection, it also can combine with naturally occurring nontoxic chemicals to form compounds that may cause cancer. The results are called *reaction products*.

Specifically, the US Environmental Protection Agency (USEPA) calls them *disinfection by-products* (DBPs) because they are formed by the disinfectants. USEPA recently set stricter limits on DBPs. Although drinking water treatment is changing to avoid the problem of reaction products, disinfection must remain adequate to kill the germs found in water. Any harmful effects to humans from DBPs are very small and difficult to measure in comparison with the risks associated with inadequate disinfection.

DBPs regulated by USEPA include a group of four chemicals with the general name **trihalomethanes** (THMs). Another group includes five **haloacetic** acids (HAA5). If you have a problem with these reaction products in your water, you will be notified by your supplier. If you're concerned, you can call them and ask whether the total THM or HAA5 level in your water is okay.

Many organic chemicals contain chlorine; one type of cleaning fluid is made up of carbon combined with chlorine for example, and many of these types of chemicals are considered to cause cancer. But not all chlorine-containing organic compounds cause cancer, so just adding chlorine to organic compounds does not always create a carcinogen.

(See Questions 26, 57, and 189 for related information.)

71. Should I be concerned about the chlorine in the water I use for bathing or showering?

No, for two reasons: (1) it will not be absorbed into the skin and get into your body; and (2) the amount of chlorine in the water is too low to harm the skin itself. There have not been any reports of danger from breathing the chlorine that gets into the air during a shower.

However, there are some people who seem to be allergic to chlorine and related compounds. This has been a problem in swimming pools. Whether this problem is caused by chlorine or chlorine reaction products is not known. If you have any trouble in swimming pools, remember that the amount of chlorine in swimming pool water is much greater than in tap water.

72. If chlorine has problems, why can't something else be used to kill the germs in water?

There are alternatives, but each one of them has problems as well. Ozone is popular in Europe. It is very strong and has no taste, but it goes away quickly and doesn't provide protection as the water comes to your home. Chlorine-related chemicals like **chloramine** and chlorine dioxide are possibilities, but chloramine is weak, so a lot must be used, and chlorine dioxide converts to toxic products, so only small amounts can be used.

Ultraviolet light is effective in clear water (the water must register 5 or less on the clarity scale and have an ultraviolet transmittance of 90 percent or greater) and under proper conditions will kill *Cryptosporidium* **oocysts,** *Giardia* cysts, bacteria, and viruses. Ultraviolet light does not form reaction products. When a disinfectant is required to be present in the water all the way to the household tap (*residual disinfectant*), chlorine or chloramine must be added afterward. There is no perfect solution. Chlorine has been the workhorse since it was first used in this country about 1900.

About 75 percent of the larger systems and 95 percent of the smaller systems use chlorine, most of the rest use chloramines. A few use ozone, chlorine dioxide, and ultraviolet light and their use may grow in the future. You can call your supplier and ask what disinfectant is used in your system.

(See Questions 26 and 82 for related information.)

Aluminum

73. I hear aluminum is used to treat drinking water. Is this a problem? Does it cause Alzheimer's disease?

Aluminum-containing chemicals—called **alum** or aluminum sulfate—are used to treat most surface waters (see Question 170). These chemicals trap dirt and then form large particles in the water that settle out; thus, very little aluminum stays in the water. Considerable publicity was given to some studies suggesting that more people got Alzheimer's disease in areas where drinking water contained small amounts of aluminum. According to most Alzheimer's disease experts, these reports are not accurate. Aluminum is a natural chemical that occurs in many foods, including tea; even if you live in areas where the level of aluminum in drinking water is much above average, your intake from food would be about 20 times your intake from drinking water. Aluminum is not regulated in drinking water in the United States.

California set a standard for aluminum in 1988 of 1.0 mg/L (see Question 188 for definition), and Arizona and Maine have established guidelines of 0.073 mg/L and 1.43 mg/L, respectively.

Health Canada has established an operational guidance value of 0.1 mg/L for water systems using aluminum-based chemicals to prevent white materials from forming in the distribution pipes. The guidance value is 0.2 mg/L for other systems.

(See Questions 19 and 170 for related information.)

74. Is it safe to cook with aluminum pans?

Yes. No health problems resulting from cooking with aluminum pans have been proven.

Radon

75. What is radon and is it harmful in drinking water?

Radon is a radioactive gas that is dissolved in some groundwaters. It is formed when radium or uranium decays naturally. When inhaled over long periods of time, radon can cause cancer. Experts also think radon has some harmful effects when consumed. When drinking water containing radon is used in your home, some of the radon goes into the air you breathe and the rest remains in the water. The US Environmental Protection Agency expects to set standards for radon in the near future, and treatment methods are already available for use by water suppliers if radon removal is required. More radon in air comes from the ground than from drinking water. Testing the air will not determine whether any radon present is seeping out of the ground or is coming from the drinking water.

(See Question 184 for related information.)

76. I'm worried that my drinking water has radon in it and that the radon will get into the air in my home. How can I test the air in my home for radon?

Contact your local health department to find out if it knows whether your drinking water contains radon or if radon is a problem in your area. This will help you decide whether you want to test for radon. The US Environmental Protection Agency (USEPA) recommends that all living areas below the third floor be tested, regardless of location, because its sampling has uncovered  high radon levels even in "low-risk" areas.

If you decide that your home should be tested, you can purchase a testing device. Two types of devices are available from a hardware and similar stores. Both contain prepaid mailing envelopes for sending the detector to a laboratory for testing. The test results then are sent back to you.

One device uses an activated carbon detector. The detector is exposed for several days and then sent to a laboratory. The activated carbon traps radon particles so they can be counted. The advantage of this device is that you obtain an answer quickly, often important when buying or selling a home.

The other device uses a film. The radon particles make weak spots in the film. At the laboratory the film is etched to form holes that are counted. This device can be exposed for periods of up to one year. Because radon levels change with time, this long-term testing is an advantage.

Radon gas seeping into a home from underground is a major source of radon in indoor air. Consequently, either type of testing device should be placed in the basement or on the first floor (if your home has no basement) where radon concentrations are likely to be highest.

Two good references are available from USEPA: *A Citizen's Guide to Radon,* 3rd edition (402-K-92-001) and *A Home Buyer's and Seller's Guide to Radon* (402-K-00-008). Single copies of both may be obtained free of charge by calling (800) 490-9198.

77. Will a water softener take radon and radium out of my water?

Radon, no. Radium, yes.

Water softeners are very effective in removing radium, but they do not remove radon. In fact, the radium that is trapped in the water softener will create some radon. Discuss this issue with your water supplier.

Remember that radium is a radioactive element that occurs naturally in some soils and thus in some groundwaters, along with radon.

(See Question 102 for related information.)

Arsenic

78. Should I be concerned about arsenic in my drinking water?

The US Environmental Protection Agency's current maximum allowable level for arsenic in community water supplies is 50 micrograms per liter (mg/L—See Question 188 for definition). The current standard is being reviewed and will likely be lowered to 10–20 mg/L by January 2002. Because most community water supplies meet the current limit, consumer's do not need to be concerned at this time (2001). If arsenic does exceed the current limit in any supply, a utility is required to notify its customers. Arsenic has been

found in some well water supplies as high as 100–500 mg/L. Users of private wells should check with their local health department about arsenic levels in the area, particularly if the region has a history of high arsenic amounts in the groundwater.

(See Questions 10, 57, and 80 for related information.)

79. If arsenic is in my drinking water, where did it come from and how did it get into the water?

Several types of mineral deposits in our natural environment contain high levels of arsenic. Groundwater flowing through these deposits can dissolve the arsenic, resulting in elevated amounts of arsenic in well water. If you use a private well, you should consider having it tested for arsenic.

(See Question 57 for related information.)

80. I have heard that arsenic is often found in well water. I have a well. Should I have my water tested for arsenic? If so, where can I get it tested?

Arsenic has no taste or smell in drinking water, so the only way you can determine whether it is in your well water is by having a water sample tested. An arsenic test by a certified state or commercial laboratory usually costs between $20 and $30. A list of certified laboratories is generally available from your local or state health department. Some certified labs may also be listed in the yellow pages of your telephone book.

(See Questions 8 and 193 for related information.)

81. My water was tested for arsenic, and it was above the US Environmental Protection Agency's allowable limit. Do treatment systems exist to remove arsenic?

Reverse **osmosis** and distillation systems are effective at reducing arsenic levels in most water supplies, and several companies are in the process of producing new adsorption systems for this purpose. Water softeners and activated carbon (sometimes mistakenly called activated charcoal or just charcoal) filters do not remove arsenic. Before installing any treatment system, consumers should thoroughly investigate the system's ability to remove arsenic, particularly if it will be used for water from a private well.

(See Question 96 for related information.)

2
Aesthetics

Yes, as everyone knows, meditation and water are wedded forever.

—H. Melville, *Moby Dick*

TASTE AND ODOR

82. Why does my drinking water taste or smell "funny"? Will this smelly water make me sick?

The four most common reasons for bad tasting or smelling water are:

- A noticeable taste can come from the chlorine that is added to the water to kill germs. Heavily chlorinated water may contain "reaction products." These products cause no taste and odor and are limited by the US Environmental Protection Agency's rules.
- A rotten-egg odor in some groundwater is caused by a nontoxic (in small amounts), smelly chemical—hydrogen sulfide— dissolved in the water.
- As some algae, bacteria, and tiny fungi grow in surface water sources, they give off

nontoxic, smelly chemicals that can cause unpleasant tastes in drinking water. Different algae cause different tastes and odors—grassy, swampy, and pigpen, as examples—and the little fungi can cause an earthy-musty taste.

- Metallic tastes can come from copper that has dissolved from copper pipe and from iron from rusting iron pipes. Copper can cause short-term health problems like diarrhea and cramping. Iron has no effect on health.

Few of the contaminants that could affect your health can be tasted in drinking water, but heavily chlorinated water may contain "reaction by-products." There are no proven incidents of the chemicals that cause a bad taste in drinking water making people sick, but just to be extra careful, water suppliers are reviewing this possibility. You should report any sudden change in taste or smell in your drinking water to your water supplier.

(See Questions 8, 17, 26, 69–72, and 96 for related information.)

83. What can I do if my drinking water tastes "funny"?

Five suggestions are:

- Store drinking water in a closed glass container in the refrigerator (warm drinking water has more taste than cold drinking water). Although some plastic bottles are okay for storing drinking water in the refrigerator, some types of plastic will cause a taste in water. If you are having trouble, use a different kind of plastic.

- Use an electric blender to mix the drinking water for five minutes. This mixing will remove some of the bad taste but not all of it. Remember that to be smelled, the chemicals that cause the smell must leave the water, get into the air, and enter your nose. When you mix the water, you hasten the chemicals leaving the water and get rid of some of the odor-causing chemicals prior to drinking the water. Then there are fewer chemicals to smell when you do drink.

- Some people object to the chlorine taste of their drinking water. Boiling tap water for five minutes should remove most, if not all, of the chlorine. If **chloramine** is used as the disinfectant in your area, boiling the water for five minutes may not remove all of the chlorine taste. Ask your water supplier what disinfectant is used in your drinking water. Because heating and boiling water use a lot of energy and create a burn risk for children

and the elderly, many people feel that water should be boiled only during emergency conditions. If you do boil, some of the minerals in the water will be concentrated a little by the boiling; however, this should not be a problem in most cases.

- Adding 1 or 2 teaspoons of lemon juice to refrigerated drinking water may result in a pleasant-tasting drink.
- To improve the taste of the water you use for drinking and cooking, install a point-of-use water treatment product that has been tested by an independent organization following the method in the Taste and Odor Reduction portion of ANSI/NSF Standard 42. These products often contain activated carbon (sometimes mistakenly called activated charcoal or just charcoal) that can remove many taste and odor causing chemicals, including chlorine. If you plan on storing water from these devices, treat the water as a food and use clean, airtight containers and refrigerate.

If the problem is a rotten-egg odor, you may wish to consider home treatment equipment that will remove hydrogen sulfide, a nontoxic (in small amounts) but offensive chemical that causes this problem.

If you have a water softener (see Question 102) that treats both the hot and cold water, chlorine will react with the softening materials inside the softener, and the chlorine will be removed. Thus, you may not have a chlorine taste, even though chlorine is added by the water supplier.

You should report any unusual taste or odor to your water supplier.

(See Questions 17, 69–72, and 96 for related information.)

84. When I put ice cubes from my freezer in water to cool it, they make the water taste funny. Why is this?

This is a common complaint that has no single, simple explanation. Many items in a refrigerator and freezer can give off odors. Freezers usually contain packaging materials, food, and metal or plastic ice cube trays. If you have an automatic icemaker, harmless bacteria can grow in the water feed line and cause odors. Smelly chemicals being used near a freezer can even be absorbed into the ice. "Freezer smell" can even sometimes be noticed in empty metal ice cube trays. Though annoying, these "off flavors" are not harmful and can sometimes be lessened by cleaning and defrosting your freezer and ice cube trays.

APPEARANCE

85. Drinking water often looks cloudy when first taken from a faucet and then it clears up. Why is that?

The cloudy water could be caused by tiny air bubbles in the water similar to the gas bubbles in beer and carbonated soft drinks. After a while, the bubbles rise to the top and are gone. This type of cloudiness occurs more often in the winter, when the drinking water is cold.

Another cause of cloudiness in cold water comes from calcium. In certain waters, the nontoxic chemical calcium carbonate will **precipitate** when it is cold. As it is white, this precipitate can cause the water to look cloudy. In this case, however, the particles settle to the bottom (usually in about 30 minutes) in contrast to the air bubbles discussed above that rise to the top of the water fairly quickly. Water with calcium carbonate precipitate in it is perfectly safe to drink or use for cooking, though it may be unappealing to look at.

86. My drinking water is reddish or brown. What causes this?

This reddish-brown color is nontoxic, but it is not harmless. It can stain clothing in the wash, and it looks bad.

The three possible causes are:

- Your drinking water may contain a brown chemical that results from the source water flowing over tree leaves, similar to the way water changes color after tea leaves are added to it. This color must be removed by the treatment plant; you can't do much about it yourself.

- Iron, which sometimes occurs in nature, may be dissolved in your drinking water. When iron is dissolved in groundwater, it is colorless, but when it combines with air as you take water from your faucet or elsewhere in the system, the iron turns reddish-brown. If you notice the water changing from colorless to brown, you may want to consider buying an iron-removal unit for your home.

- Drinking water pipes—in the street, leading to your home, or in your home—may be rusting, creating rusty-brown water. Also, your hot water tank may be rusting. Water causing this type of problem is called corrosive. If you are having trouble and your neighbors are not, then your own pipes or water heater probably are rusting. Letting the water run a while will often clear the water (save the rusty water for plants). When your plumbing is rusting, lead and copper may be getting into your drinking water as well. This is important, so call your local water supplier to discuss this. To avoid problems with lead and copper, all water suppliers by law have to make sure that drinking water is not corrosive.

(See Questions 61, 63, and 96 for related information.)

87. My drinking water is dark in color, nearly black. What causes this?

When manganese, a chemical currently thought to be nontoxic that frequently occurs in nature, dissolves in groundwater, it is colorless. When it combines with the chlorine in the water as it comes to your home, it turns black. To prevent "black water" problems, the US Environmental Protection Agency established a recommended limit (not required, just recommended) for manganese in drinking water. If you have blackish water, you may want to consider a filter to remove manganese from the water in your home. You should also report your problem to your water supplier.

(See Question 96 for related information.)

3
Home Facts

Water, which so many townspeople never think about, having an obedient spring in the kitchen, is really among the most fragile of life's necessities.

—H. V. Morton, *The Waters of Rome*

GENERAL

88. How long can I store drinking water?

Drinking water that is thoroughly disinfected, such as water from your public water supplier, can be stored for six months in capped, plastic containers that will not rust. Glass containers should be avoided as they can easily be broken. Water that has been boiled for one minute, or three minutes at high altitudes, can be stored for up to one year. Be sure to cool the water before storing it. Be careful to use plastic that will not make the water taste bad—trial and error is best here. Because the disinfectant that was in the water when you stored it will slowly go away, replacing the water every six months is recommended. The taste will become "flat" after extended storage, so periodic replacement will help here also. If possible, you should store water in a refrigerator to help control bacterial (not germ) growth.

(See Question 112 for related information.)

89. How much water should I store for emergencies?

A good rule of thumb is to store one gallon of water per person per day. Plan for at least three days. For example, a family of four should store about 12 gallons of water. You'll need more water in hot temperatures and for strenuous activities. People with special needs such as nursing mothers, young children, and family members with illnesses may require more water.

(See Question 88 for related information.)

90. Is it okay to use hot water from the tap for cooking?

No. Use cold water. Hot water is more likely to contain rust, copper, and lead from your household plumbing and water heater because these contaminants generally dissolve into hot water from the plumbing faster than into cold water.

While we're on
the subject of hot
water, here's a good
conservation tip:
Insulating your hot
water pipes will
help keep the water
in them warm
between hot water

uses. Thus, after the first use of the day, hot water
will come to the tap sooner, conserving water.

(See Questions 61 and 63, and the Chapter on
Conservation for related information.)

91. Is it okay to use hot water from the tap to make baby formula?

No. As noted in Question 90, hot water may
contain impurities that come from the hot water
heater and plumbing in your house. To avoid this,
use cold water and let the water run for a couple
of minutes before you use it if that tap has not
been used for a while, overnight, or all day. You
can then heat this water in the microwave or on
the stove. Catching the water you flush out of the
tap in a container and saving it for plant watering
is a good conservation measure.

92. Is it okay to heat water for coffee or tea in a microwave in a styrofoam cup?

Yes. The chemicals in the styrofoam are not affected by the microwaves and do not "melt" and get into the water.

93. Is it okay to drink water that comes out of a dehumidifier?

No. Don't drink this water because it has not been disinfected. This water is fine for watering plants (add fertilizer as needed), car batteries, or steam irons.

94. What should I do to avoid cold-weather problems with my pipes?

There are several necessary steps homeowners should take to avoid freezing pipes. First, disconnect and drain outdoor hoses. Detaching the hose allows water to drain from the pipe. Next, insulate pipes or faucets in unheated areas. Make sure you locate your master valve in case pipes freeze and rupture. Also, check with your local water company; you may be responsible for keeping the meter from freezing as well. In other places, the meters must be maintained by utility personnel. If you are expecting severe cold weather and are worried about your pipes freezing, you can also leave a *small* stream of water flowing in the bathroom during the worst of the cold spell. Let just enough of water through to produce a steady, thin stream. Any extra flow is a severe waste of water.

(See Question 95 for related information.)

95. How can I locate my home's master valve?

It is important to know where the master valve is in case you have a major leak. The most common locations in your house or apartment are:

- Where the water supply enters your home
- Near your clothes-washer hook-up
- Near your water heater

To determine if the valve you have found is the correct one, try turning it off and see if it shuts off all water faucets in your home. If not, repeat this process with each valve you find until you identify the correct one. If you are unable to locate it, contact your plumber for assistance. Once you've found the valve, it's a good idea to mark it with something distinctive like bright paint, a tag, or ribbon. This will help you locate it quickly in case of an emergency.

TREATMENT

96. Should I install home water treatment equipment?

The US Environmental Protection Agency has indicated that point-of-use (POU) and point-of-entry (POE) home water treatment devices can be used to meet requirements of the Safe Drinking Water Act regulations for some contaminants. In these cases, your water supplier must provide the device as part of the utility's overall responsibility for ensuring wholesome drinking.

Otherwise, this is a personal decision. This type of equipment is generally not needed to make your water meet federal, state, or provincial drinking water standards. If you decide to use a home water treatment device, you must be careful to maintain it properly, according to the manufacturer's instructions. For example, modern POU/POE treatment products often come equipped with a yellow and red light monitoring system, or the water flow is automatically shut off to warn users when to change replaceable cartridges and perform routine maintenance.

If you are concerned about additional aesthetic qualities, like tastes and odors, or providing additional barriers against chlorine-reaction products and residual traces of contaminants such as pesticides, solvents, lead, or *Giardia* cysts and *Cryptosporidium* **oocysts,** you might consider a home treatment unit. POU systems can be located in several places in the home: countertop, faucet-mounted, under-sink cold tap, or under-sink line bypass. POE systems are located where the water comes into the house.

Treatment units can be grouped into six general categories:

- Particulate filters that remove particles, including black manganese particles, of different sizes.

- Adsorption filters (most of which are not really filters) usually contain activated carbon (sometimes incorrectly called activated charcoal or just charcoal) that remove chlorine, tastes and odors, and organic compounds like pesticides. Some units are capable of removing chlorine-reaction products and solvents such as cleaning fluid. Harmless microbes do grow in these units, but these are not germs. Most adsorption filters remove very little copper and lead. Certain special filters will remove dissolved lead, but check their claims with independent organizations, as noted at the end of this answer.

- Oxidation/filtration systems that will change iron (clear water turning red) or hydrogen sulfide (the rotten-egg odor) into a form where these nontoxic but troublesome chemicals can be filtered out of the water before it comes into your home. These are frequently used by people who have their own source of water, such as a private well.

- Water-softening systems can trade (exchange) nontoxic chemicals that cause "hardness" for other nontoxic chemicals that do not cause hardness. These units must be renewed (regenerated) periodically with salt.

- Reverse **osmosis** units that remove hardness; chemicals such as nitrates, sodium, dissolved metals (such as lead and copper) and other minerals; and some organic chemicals. Reverse osmosis units also remove fluoride. Some units are sensitive to chlorine, so a chlorine-removal step usually is included prior to the reverse osmosis unit. Reverse osmosis units do allow some organic chemicals to pass into the treated water, however. Therefore, sometimes these systems are followed by adsorption units to remove these organic compounds. Reverse osmosis units usually produce relatively small volumes of water.

- Distillation units that boil the water and condense the steam to create distilled water remove some organic and inorganic chemicals (hardness, nitrates, chlorine, sodium, dissolved metals, and so forth). Distillation units also remove fluoride. However, some organic chemicals may pass through the units with the steam and contaminate the distilled water unless the unit is specifically designed to avoid this problem.

All of these units require maintenance and should be bought from a reputable dealer. Their performance should be tested and validated against accepted standards like those used by NSF International, the Water Quality Association, and Underwriters Laboratories, Inc. These standards allow manufacturers' performance claims for "drinking water treatment unit" (DWTU) products to be tested and checked. You should investigate all claims made for any unit by visiting the Internet site of a testing organization or by calling the organization directly. NSF International: <http://www.nsf.org/Certified/DWTU/>, (800) 673-8010, Water Quality Association: <http://www.wqa.org/goldseal/>, (800) 749-0234, Underwriters Laboratories, Inc. <http://database.ul.com/cgi-bin/XYV/template/LISEXT/1FRAME/ccnsrch.html>, type FDQD (all caps) in the category control number box and click on enter, (847) 272-8800.

Remember that if the equipment removes the disinfectant in your tap water, the treated water should be treated as a food if it is stored.

(See Questions 59, 64, 81, 83, 86, 87, 100–102, and 186 and Appendix A for related information.)

97. I heard about a water treatment device that uses an electromagnet to treat water. Does this work?

The Water Quality Association (WQA), a not-for-profit organization representing the household, commercial, and light industrial water treatment industry, issued a "Magnetics Task Force Report" in 2001. The report concludes there is not sufficient scientific evidence to determine the effectiveness of physical water treatment technologies such as magnetic, electromagnetic, and **catalytic** devices in home water treatment devices without proof of performance claims through testing according to methods of an American National Standards Institute "Drinking Water Treatment Unit" standard for magnetic and other physical water treatment products. At this time (2001), no such standard exists.

98. How do I treat my water if my supply fails?

If you need to disinfect your water, you have several options. As mentioned in Question 28, the first method is to boil it. Bring the water to a boil, boil it for one full minute (three minutes if you live at a high altitude), then allow it to cool before storing (see Question 88 for storing information). Caution is required because heating and boiling are burn hazards.

When boiling is not practical, you should use chemical disinfection. Two commonly used chemical disinfectants are chlorine and iodine. They are somewhat effective in protecting against *Giardia* **cysts** but may not be effective against *Cryptosporidium oocysts*. Therefore, use chlorine or iodine only on groundwater supplies (wells), which are not likely to have these germs. Water from rivers, lakes, reservoirs, or springs should be boiled.

Common household bleach contains a chlorine compound that will disinfect well water. Check the label: if the available chlorine is listed as around 1 percent, add 10 drops to one quart of water; if 4–6 percent, add 2 drops to one quart of water; if 7–10 percent, add one drop to one quart of water. If the percentage is not listed, add 10 drops to one quart of water. Double the number of drops if the water is colored or cloudy. Mix the water thoroughly and allow it to stand for 30 minutes. After this time, the water should have a slight chlorine odor; if not, repeat the chlorine addition and allow the water to stand an additional 15 minutes. If the chlorine taste and odor is too strong, let it stand exposed to the air for several hours or pour it back and forth from one clean container to another.

Common household iodine from a medicine cabinet or first aid kit may also be used. Add five drops of 2 percent U.S.P. tincture of iodine to one quart of well water. If the water is colored or cloudy, add ten drops. Let the water stand at least 30 minutes before use.

99. How can I protect my private water supply?

You can protect a private water supply by carefully managing activities near the water source. For households using a domestic well, this includes keeping contaminants away from sinkholes and the well itself. You should also keep hazardous chemicals out of septic systems. Technical assistance is available to help you protect your water supply. The National Drinking Water Clearing House (800-624-8301, <http://www.ndwc.wvu.edu>) and the organization Farm*A*Syst/Home*A*Syst (608-262-0024, <http://www.uwex.edu/homeasyst/>) can provide helpful information.

(See Question 117 for related information.)

HARD AND SOFT WATER

100. Is distilled water the "perfect" drinking water?

Distilled water has advantages and disadvantages.

Distilling removes many potentially harmful chemicals like lead, copper, nitrates, sodium, some organic contaminants, and chlorine. Boiling water to make distilled water will kill any harmful bacteria and viruses, as well as *Giardia cysts* and *Cryptosporidium oocysts*. Distilling removes beneficial fluoride, and some organic contaminants like chloroform and cleaning fluid (solvents) may leave the water with the steam and end up in the final water when the steam is cooled. However, most companies that provide water distillers incorporate additional treatment into the system to prevent any organics carried with the steam from ending up in  the final product. Finally, because most of the minerals are missing, using distilled water in a kettle to make tea or coffee will avoid the buildup of scale (the white stuff).

Except in special cases for taking salt out of sea water to make drinking water, distilled water is too expensive to be provided to your house by your public water supplier.

Although some think that the low mineral content of distilled water is a disadvantage, most people consume plenty of minerals in a well-balanced diet. Finally, there is some disagreement over the taste of distilled water. Many people like it; others find it flat and tasteless.

Distilled water is handy around the home for use in steam irons and car batteries and for watering plants.

(See Questions 59, 96, and 125 for related information.)

101. What is "hard" water?

"Hardness" in drinking water is caused by two nontoxic chemicals (usually called *minerals*)— calcium and magnesium. If calcium and/or magnesium is present in your water in substantial amounts, the water is said to be *hard* because making a lather or suds for washing is *hard* (difficult) to do. Thus, cleaning with hard water is hard/difficult. Water containing little calcium or magnesium is called *soft* water. (Maybe it should be called easy, the opposite of difficult.)

(See Questions 63, 96, and 175 for related information.)

102. Should I install a water softener in my home?

If you are bothered by a soap deposit in your bathtub or by the buildup of white deposits (called scale) on your cooking pots and coffee maker, a water softener can help with these  problems. You can find out the hardness of your drinking water by telephoning your water supplier. The higher the hardness number, the more a water softener will help. A water softener can reduce the formation of scale in your hot water system and make washing easier.

The water softener replaces the nontoxic "hardness" minerals with sodium or **potassium**. The amounts of these elements are relatively insignificant in comparison to what you get in food and should not be a problem, unless your doctor has put you on a special restricted diet.

Whether to put the softener on your main water line or just the hot water line is a complicated issue. Softening only the hot water has some cost and environmental advantages related to *regeneration*, which is a process by which the softening materials (called *resins*) inside the softener can be used over and over again.

Water softeners are regenerated with salt. After the salt is used, it goes down the drain and into the environment—so the less salt used the better. Using less salt also saves you money. If you soften only the hot water, less water goes through the softener, so it needs regeneration less often, meaning less salt is being used. Also, regenerating a softener after a selected amount of water has gone through it rather than on a particular time schedule is better, as this prevents wasting salt by regenerating too soon or using the softener after it has stopped softening.

Finally, some people think bathing in completely soft water (both hot and cold water softened) is unpleasant—it feels like the soap won't rinse off. You may be surprised to learn, however, that rinsing is actually more complete in soft water than in hard water. Although you can't see it, when you bathe or wash your hair in hard water, some of the same stuff that causes the bathtub ring gets on your body or in your hair. With soft water this material does not form, so rinsing is more complete.

Softening only the hot water has two disadvantages. First, if you wash your clothes in cold water, you won't get the benefit of soft water; however, you can buy products to add to your wash to help if this is a problem. Second and more important, if your water is very hard when you mix the hot and cold water together, the water will still be hard and you won't see much benefit from the softener.

Some people have expressed concern about whether the installation of a water softener might raise the lead and copper content of drinking water in homes that are experiencing problems. Probably not. In April 1998, the US Environmental Protection Agency completed a research report (EPA/600/R-98/044) concluding that "Considering the metal leaching [dissolving into the water] as a whole, the ion exchange softened water did not show a pattern of higher metal leaching from plumbing materials. The test results suggested, therefore, that ion exchange softening will not necessarily increase the **corrosivity** [dissolving of metal into water] of water."

(See Questions 61, 63, 77, 103, and 104 for related information.)

103. Does drinking water that has been softened deplete the body of calcium and magnesium?

No. Drinking softened water does not drain calcium and magnesium from the body.

104. I have a water softener, but I still get spots on my bathroom tile. Why is this?

First, a little easy chemistry. All water contains dissolved nontoxic minerals. Calcium, magnesium, sodium, sulfate, chloride, and bicarbonate are the most common. These have no health effects. If you put water in a dish and let it evaporate, the white stuff that is left are these minerals.

Two of these minerals, calcium and magnesium, cause "hardness." Because they interfere with sudsing and thus make washing "hard" (difficult), they are referred to as causing hardness. A water softener like you have trades (exchanges) the calcium and magnesium for sodium or **potassium**, so the water leaving the softener has no calcium and magnesium (thus, no hardness) but more sodium or postassium.

Thus, minerals were not eliminated during softening, just traded. If you put the softened water in a dish and let it evaporate, the white stuff left over, although it would be different, would look the same and would equal the same amount as before the softener was installed.

If you had installed a reverse **osmosis** unit instead of a water softener, it would actually remove minerals from the water, not just exchange them, and you would not experience this problem.

(See Questions 101 and 102 for related information.)

105. When I put ice cubes that I've made in my freezer into a glass of water, white stuff appears in the glass as the ice cubes melt. What is the white stuff and where does it come from?

Ice cubes freeze from the outside, so the center of the cube is the last to freeze. Ice is pure water, only H_2O, so as the ice cube freezes, all of the dissolved minerals, like the hardness minerals (see Question 101), are pushed to the center. Near the end of the freezing, there isn't much water left in the center of  the cube, so these minerals become very concentrated, and they form the "white stuff"— the technical name is **precipitate**. The hardness minerals that cause the "white stuff" are not toxic.

Some commercial ice cubes are "cored" after they freeze to remove this material. Having posts in your ice cube tray doesn't help, however, as the precipitate must actually be removed by coring.

106. What is that white stuff in my coffee pot and on my showerhead and glass shower door? How can I get rid of it?

Minerals dissolved in water tend to settle out when water is heated or are left behind when it evaporates. These minerals are white and accumulate in coffee pots and on showerheads and glass shower doors.

To remove these minerals, fill the coffee pot with vinegar and let it sit overnight, or soak the showerhead overnight in a plastic bowl filled with vinegar. Be careful not to spill this mixture. When you are done, carefully discard the contents of the plastic bowl down a drain, and flush the container and sink drain with plenty of water. *NOTE: Rinse the coffee pot or showerhead thoroughly after treatment and before use.* Pouring the excess hot liquid out of your coffee pot when you are finished with it will help somewhat in preventing this problem.

White spots on glass shower doors are difficult to remove with vinegar because the spots dissolve very slowly. A better idea is to prevent the spots from forming by wiping the glass door with a damp sponge or towel after each shower.

NOTE: Some commercial establishments use untreated water for irrigation to save on tap water. If this is groundwater, it may be high in minerals (treatment beyond that mentioned in Question 170 is needed to get minerals out of the water) and if this water sprays onto your car, it can leave spots. Vinegar will remove them. Rinse with good water after using the vinegar.

107. Why does my dishwater leave spots on my glasses?

The spots that may appear on glassware after it is washed and air-dried are caused by nontoxic minerals that remain on the glass when the water evaporates.

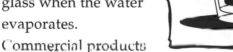

Commercial products are available that allow the water to drain from the glassware more completely. As noted in Question 106, spots on glass shower doors appear for the same reason.

108. Some of my clear glassware comes out of the dishwasher with a rainbow sheen on it. What causes this?

Too much dishwasher detergent can leave an oily looking sheen on glassware. Also, *Consumer Reports* has noted that some detergents can actually etch glass. The rainbow coating might be a fine pattern of etching caused by the detergent.

109. What causes the whitish layer on the soil of my potted plants?

Drinking water can contain many nontoxic chemicals. When the water on your plants evaporates, these chemicals are left behind as a whitish layer. Using distilled water on your plants will avoid this problem. Catching rainwater for watering plants is a good idea but be sure to cover the water so that mosquito larvae do not grow in it, unless you bring it in after each rain. If you have a dehumidifier, the water that comes out of it is good for plant watering. Distilled water, rainwater, and dehumidifier water will wash minerals out of the soil, however, so use a slow-release balanced fertilizer to replace them.

(See Questions 93 and 125 for related information.)

BOTTLED WATER

110. Should I buy bottled water?

You don't need to buy bottled water for health reasons if your drinking water meets all of the federal, state, or provincial drinking water standards (ask your local supplier). If you want a drink with a different taste, you can buy bottled water, but it costs up to 1,000 times more than municipal drinking water. Of course, in emergencies bottled water can be a vital source of drinking water for people without water.

The US Food and Drug Administration (FDA) requires bottled water quality standards to be equal to those of the US Environmental Protection Agency for tap water, but the quality of the finished product is not government-monitored. Bottlers must test their source water and finished product once a year. Currently, any bottled water that contains contaminants in excess of the allowable level is considered mislabeled unless it has a statement of substandard quality. Regulations require bottlers to inform consumers of "bottled water" contents. Although recent tests have not found any lead in dozens of brands of bottled water, studies have shown that microbes may grow in the bottles while on grocers' shelves. Some states impose expiration dates on bottled water, two years from the date of bottling in New York, for example. Canada does have restrictions on labeling bottled water and has minimal quality requirements covered by the Canadian Food and Drug Act.

In May 2001, a Swiss-based conservation group, World Wide Fund for Nature, reported that, "Bottled water may be no safer or healthier than tap water in many countries while it sells for up to 1,000 times the price." The group also noted that 1.5 million tons of plastic are used each year to bottle water. The International Bottled Water Association responded: "If municipal water is used as a source for bottled water, it typically undergoes additional processing and purification for quality, safety, and taste." This association also noted that the bottled water industry is preparing to adopt worldwide standards in summer 2001 to ensure uniform quality.

Bottled water is popular; Americans spend about $3 billion annually to buy this product—half the amount the country spends to protect tap water. Overall about 15 percent of US households drink bottled water regularly. Remember, if you use bottled water, consider it a food and refrigerate it after opening.

NOTE: Individuals placed on a highly restricted sodium diet should choose a brand of bottled water that contains zero (0) milligrams (mg) of sodium in an 8-ounce glass. CAUTION: Some bottles labeled sodium-

*free contain some sodium, maybe too much for those on
a highly restricted sodium diet. Check the label carefully
on any bottle of water you buy to find out the sodium
content of that particular brand, regardless of the
general labeling.*

(See Questions 42 and 186 for related
information.)

111. What do the labels on bottled water mean?

Effective May 13, 1996, the Food and Drug
Administration (FDA) established new rules for
the labeling of bottled water.

Artesian water* or *artesian well water is water
that comes from a well drilled into a confined
aquifer in which the water level stands at some
height above the top of the aquifer.

Groundwater is water from a subsurface
saturated zone that is under a pressure equal to or
greater than atmospheric pressure. Such
groundwater must not be under the direct
influence of surface water.

Mineral water is water that contains not less
than 250 milligrams per liter (mg/L) (see Question
188 for definition of mg/L) of total dissolved
solids (TDS—determined by evaporation to
dryness and weighing the residue) coming from a
source tapped at one or more bore holes or
springs, originating from a geologically and
physically protected underground water source.

Purified water or *demineralized* water is water that has been produced by distillation, deionization, reverse **osmosis**, or other suitable

process that meets the definition of purified water in the *United States Pharmacopeia*, 23rd revision, January 1, 1995. Alternatively, water may be called *deionized water* if the water has been processed by deionization, *distilled water* if the water has been processed by distillation, *reverse osmosis water* if the water has been processed by reverse osmosis, and so forth.

Sparkling bottled water is water that, after treatment and possible replacement of carbon dioxide, contains the same amount of carbon dioxide that it had as it was taken from the source.

Spring water is collected from an underground formation from which water flows naturally to the surface of the earth. Water must be collected at the spring or through a bore hole tapping the underground formation feeding the spring. A natural force must cause the water to flow to the surface and the location of the spring must be identified.

Sterile water or *sterilized water* is water that meets the requirements of the sterility tests in the *United States Pharmacopeia*, 23rd revision, January 1, 1995.

Well water is water from a hole, bored, drilled, or otherwise constructed in the ground that taps the water of an aquifer.

The FDA requires the following additional labeling requirements:

If the TDS content of mineral water is below 500 mg/L or above 1,500 mg/L, the statement *low mineral content* or *high mineral content* must be added to the label, respectively.

If the source of the bottled water is a community water supply, *from a community water system* or *from a municipal source* must be added to the label.

If the product states or implies that it is to be used for feeding infants and the product is not commercially sterile, *not sterile* must be added to the label.

Except for bottled water described as *water, carbonated water, disinfected water, filtered water, seltzer water, soda water, sparkling water*, and *tonic water*, the FDA has mandated a list of microbial, physical, and chemical tests that cover the quality of these products. They are similar to the quality standards the US Environmental Protection Agency applies to tap water. If the product does not meet these standards, the terms *contains excessive bacteria, excessively turbid, abnormal color, abnormal odor, contains excessive chemical substances* (unless it is mineral water, then *contains excessive [specific chemical]* is used), or *contains excessive radioactivity.*

No sampling frequency or enforcement requirements are included in the regulations.

112. I want to store some water for a possible emergency. Is bottled water okay to store?

No. Bottled water is a good source of drinking water during emergencies and when a person is on the go, but it does not store well. Many water bottlers commonly disinfect bottled water with ultraviolet light and also add ozone, but because these methods don't leave any disinfectant in the water, microbes can grow over time. These microbes are not germs, so they won't make you sick, but to maintain freshness, the International Bottled Water Association recommends that water bottles be labeled with the bottling date and be replaced every six months. As mentioned in the answer to Question 110, bottled water has a shelf life in a store as well.

As noted in Question 88, tap water, which does contain a chemical disinfectant, should be stored in proper containers for an emergency, although even it will not last indefinitely.

If your water stops during an emergency, remember the water in your hot water-tank, melted ice cubes, and the water in your toilet tank reservoir can be used. If you have the ability to do so, boiling these sources of water is always a good idea before drinking.

(See Questions 88 and 113 for related information.)

113. Is it okay for people who sweat a lot—for example, athletes, outdoor workers, and dancers— to carry bottles of water with them?

Yes, fluid replacement is extremely important if you are sweating a lot. Tap water carried in clean bottles is a good source of something to drink when you're on the go. Just be sure to change the water often, at least daily. Also, avoid storage in a hot car. The warm temperature can cause any microbes in the water to grow quickly.

114. Should I buy drinking water from a vending machine?

Buying water from a vending machine is a matter of personal choice. The treatment of vending machine water is based on sound scientific principles. Treatment such as reverse osmosis, activated carbon adsorption, and ultraviolet light disinfection are often used in vending machines. Such treatment (beyond typical water treatment) can enhance the quality of many water supplies, going beyond the safety level specified by federal, state, and provincial authorities. As with any mechanical equipment, vending machine treatment devices must be regularly maintained and water quality tested to provide satisfactory operation. If you use a machine in which you have confidence, be sure to put the water in very clean bottles, refrigerate when you get it home, and treat it like a food.

OTHER USES IN THE HOME

115. How much water does one person use each day?

Total water use varies depending on lawn watering, if any, and whether a home has a washing machine and dishwasher. The US average is nearly 50 gallons used each day by each person. Of this, the amount used for cooking and drinking varies among individuals, from about 13 ounces to about 2 quarts. The average use is about 2.5 pints—about half for plain water consumed as a beverage and the rest consumed in other beverages (juice, coffee, and so forth) and used for cooking.

Because of other uses in a community, the water supplier pumps much more water than is just used in households. A study of 1,100 water suppliers around the United States showed that to supply all the water needed for all uses, the average amount of water pumped was 180 gallons each day for each person.

In Canada, average total home water use is about 60 US gallons for each person each day. The use for drinking only has been estimated at about 1.5 liters (1.6 quarts) each day.

(See Questions 49 and 130 for related information.)

116. Where does the water go when it goes down the drain?

If you are on a sewer system, all of the drains in your house are connected to a single pipe that leads to the street. The pipe in the street collects the wastewater from all the homes in your area and takes it to a larger pipe that collects water from other streets. The wastewater then flows into still bigger pipes that connect various neighborhoods. Think of a large tree with your house at the tip of a branch near the top. Like the tree branches that are bigger nearer the ground, the pipes in the wastewater collection system are larger and contain more liquid as they near the wastewater treatment plant. Here, the wastewater is treated and cleaned so that it can be put back into the environment without harming anything. A drinking water distribution system (see Chapter 7) looks the same but in this case the drinking water goes from the treatment plant to your home.

If you are not connected to a sewer system, the liquid wastes from your home go into a septic tank, where most of the solids settle out. The water then goes into a leach field, pipes buried in the ground that have holes in the bottom. The water seeps out of these holes and into the ground.

117. What can I safely pour down the sink or into the toilet?

Before you think about what you can throw away, think about what you are buying. Start by buying environmentally friendly products whenever possible.

Next, try to buy just what you'll need so you won't have any or very much left over. Finally, check with your local department of solid waste or similar department for local rules and find out if there are hazardous waste collection days.

If your home is on a sewage system, these liquids can safely be poured down a drain, followed by *plenty* of flush water:

- Aluminum cleaners
- Ammonia-based cleaners
- Drain cleaners
- Window cleaners
- Alcohol-based lotions
- Bathroom cleaners
- Depilatories
- Hair relaxers
- Medicines (expired)
- Permanent lotions
- Toilet bowl cleaners
- Tub and tile cleaners
- Water-based glues
- Paintbrush cleaners with trisodium phosphate
- Lye-based paint strippers

After disposal, be sure to rinse the empty container with water several times. Of course, the safest course of action is not to put anything in your sink or toilet.

In Canada, the local sewer-use bylaw controls disposal in most municipalities.

118. I have a septic tank. Should I take any special precautions?

If you have a septic tank, checking with your local authorities is a good idea, but here are some general rules. First, remember that any substance you put down the drain into a septic may eventually get into the local groundwater. Second, don't bother with septic-tank additives or the addition of yeast; they really don't help the septic tank very much. Third, follow these recommendations:

- Do not dispose of fats, grease, or cooking oil down the drain.
- Do not use a garbage disposal or put coffee grounds, meat bones, or other food products that are difficult to biodegrade down the drain.
- Do not dispose of household cleaning fluids down the drain and use disinfectants sparingly.
- Do not dispose of automotive fluids such as gas, oil, transmission or brake fluid, grease, or antifreeze down the drain.
- Do not dispose of or rinse any containers containing pesticides, herbicides, or other potentially toxic substances down the drain.
- Do not dispose of any nonbiodegradable substances or objects such as cigarette butts, disposable diapers, and feminine hygiene products down any drain or toilet.

Minimize water usage. Do not run water continuously while rinsing dishes or thawing frozen food products (these are good conservation measures in any household). Consider limiting toilet flushes or putting a plastic bottle full of water in the toilet tank to reduce the amount of water used in each flush (see Question 131).

Run only full loads when using a dishwasher or washing machine (again, good conservation ideas in any household). Try to use the washing machine at times when water is not being used for other purposes.

- Do not use any chemicals to clean your system, they may actually harm the system or the groundwater.
- Do not connect any footing or foundation sump pumps to the septic tank system.

See Question 117 for a list of what can be safely disposed of down the drain if your household is connected to a municipal sewer system.

119. Why do hot water heaters fail?

Because of the natural corrosive properties of all waters, holes will eventually rust through a water heater wall. The time it takes for this to happen varies depending on how corrosive your water is. To avoid problems with lead and copper, water suppliers are required by regulation to make water less corrosive.

In hard water areas (see Question 101), the minerals causing the hard water tend to form a "scale" (see Question 102) at the bottom of the hot water heater, resulting in failure of the unit. Using a water softener should help alleviate this problem.

(See Question 86 for related information.)

120. What causes the banging or popping noise that some water heaters, radiators, and pipes make?

Each noise has a different cause. In a water heater, some of the nontoxic minerals in the water form a rough coating on the inside of the heater when the water warms up. When the container walls are rough, air bubbles form before the water boils. These bubbles burst as the water is heated, causing a popping noise. In a smooth-walled container, heated water will not form bubbles until it boils, and boiling does not occur in a hot water heater, so no noises will be heard when a water heater is new. Occasionally, flushing the water heater from the bottom will prevent some but not all of the coating from forming. Using a water softener should minimize this noise.

In a home radiator heated with steam, the banging noise is caused by steam bubbles collapsing in water that is pooled in the system.

Pipes make noises for two reasons. First, when you open a hot water tap after water hasn't been used for a while, the pipe will be cold. As the hot water runs through the pipe, the pipe heats up and gets bigger. This will sometimes cause the pipe to creak or make other noises.

The other reason is *water hammer*. When water is running through a pipe fast and the flow is stopped quickly, the water keeps moving for a while, like a train plowing forward during a wreck. The moving water finally bangs against (hammers) the faucet or valve, making a loud noise, like a hammer hitting metal. If you've noticed this problem in your home, it can easily be corrected by simply turning the water off more slowly.

(See Question 102 for related information.)

121. How should I fill my fish aquarium?

First, allow at least 1 gallon (4 liters) of water to run from the tap before using the water to fill the aquarium. This will flush any copper or zinc from copper or galvanized piping in your home;

tropical fish are very sensitive to small amounts of copper or zinc in their water. Saving this water for other purposes such as plant watering is a good conservation measure. With a plate in one hand, pour water over the plate into the aquarium, allowing the water to drop about 1 foot (30 centimeters) before hitting the plate. This will add air (oxygen) to the water. Let the water sit in the aquarium for an hour or two until it reaches room temperature. Consult your local pet store to learn how to test for and remove any disinfectant in the water. Remove the disinfectant from the water in the aquarium before adding the fish.

122. I have trouble keeping fish alive in my fish pond. Is there anything I can do?

Fish get sick and die for many reasons. One problem is that waste products from the fish in a fish pond can decay and release ammonia, which is quite toxic to the fish and other aquatic life. Commercial products are available that will take up the ammonia and exchange it for a nontoxic chemical. If you use a "biological filter," the ammonia will be changed to a chemical nontoxic to fish by microbial action. Of course the disinfectant in tap water and diseases are other possible causes for fish to die. Check with your local pet store.

123. When I try to root a plant or grow flowers from a bulb in my house, the water looks terrible after a while. What will prevent that?

Put 1 to 2 tablespoons of activated carbon (sometimes called agricultural charcoal) in the bottom of the bowl. This will help keep the water in better condition. Of course, changing the water frequently will help also.

124. Roses, azaleas, camellias, and rhododendron all require acid conditions. How should I adjust the acid content of my plant water?

The acid content of water is measured by the pH. Any number below 7.0 indicates that the water is acidic. These plants like water with a pH somewhere between 6.5 and 6.8, lower than found in most tap waters. Acidic fertilizers, vinegar, or a very little muriatic acid (handle with care) can be used to lower the pH of tap water before using it on potted plants of this type. Before deciding how much acid to add to a gallon of water, do some testing using a pH test kit that can be obtained from garden stores.

125. I live in a very hard water area and I have a water softener. My plants don't seem to like my tap water. What can I do?

As noted in Question 102, ion exchangers usually replace (exchange) the hardness chemicals (calcium and magnesium) with sodium. If you soften very hard water, you will wind up with quite a bit sodium in your tap water, and some plants don't like sodium. Discuss this with your local garden store. You can try one of four things: (1) use reverse osmosis–treated water or distilled water for watering plants; (2) change the chemical you use for regenerating your softener from sodium chloride to **potassium** chloride (although this may be more expensive and harder to get); (3)

take water for the plants from a tap before the water gets to the softener, for example, from an outside garden hose connection near the street; or (4) if you are using sodium softened water on your plants, water heavily to rinse off previously deposited minerals. Heavy sodium salt concentrations in the absence of calcium and magnesium may cause the dirt to swell a bit and retard the growth of plant roots.

(See Questions 102 and 109 for related information.)

COST

126. What is the cost of the water I use in my home?

Most people pay for water delivered to their home according to the amount they use. In the United States the water rate is charged for each 1,000 gallons used (in other countries the charge is for each cubic meter used). Prices vary greatly, but a typical cost is about $2 (U.S.) for 1,000 gallons. A gallon of tap water costs less than one penny. For the cost of one bottle of designer water you could refill it 2,000 times with tap water.

One thousand gallons of water would serve one consumer for about 20 days, so tap water is not very expensive. Of the amount charged for 1,000 gallons, about $0.30 to $0.50 is for treatment; the rest is for paying the mortgage on the treatment plant and the pipes in the street, the salaries of the employees who work for the drinking water utility, and some profit for privately owned water companies.

You can figure the cost of water in your area by looking at your water bill and dividing the total cost for water by the total amount of water used (just use the water part of the bill if other costs are included). In general, in the United States we spend about 0.5 percent of our income on both drinking water and wastewater disposal.

(See Questions 115 and 164 for related information.)

127. How does the water company know how much water I use in my home?

Most households have a water meter that measures the amount of water used. In some communities, households are charged the same cost (called a flat rate) each month. For those communities with water meters, a person from the water utility reads the meter on a regular schedule. The previous reading is subtracted from the current reading to determine the amount of water actually used. In some cases, the cost of water to residents is covered by general taxes.

128. How does the water company know that my water meter is correct?

Most water companies have programs to routinely test water meters on a rotating basis to make sure the meters are accurate. Of course, if your recorded water use changes suddenly for no obvious reason (more people in the home, away for a long trip, or heavy lawn watering), report this to your water supplier so it can be investigated. In most instances, when a water meter is wrong, it reads low. As a good citizen, you should report this to your water supplier just as you would when you think your meter might be reading high.

129. We had a conservation drive in our area and everyone cooperated. Then our water rates went up. Why?

Water suppliers have fixed costs—salaries, mortgage payments, and so forth. They must collect this money regardless of water use, so when water volume goes down because of conservation by the public, the cost of each gallon of water used sometimes is raised to provide the water supplier with the money it needs to operate.

(See Chapter 4 for related information.)

4

Conservation

To take anything for granted, is in a real sense, to neglect it and that is how most of us treat water.

—Robert Raikes, *Water, Weather, and Prehistory*

130. What activity in my home uses the most water?

Toilet flushing is by far the largest single use of water in a home. Most toilets use from 4 to 6 gallons (15 to 23 liters) of water for each flush. Canadian flush toilets use about 4 to 6 imperial gallons (18 to 28 liters). On the average, a dishwasher uses about 50 percent less water than the amount used when you wash and rinse dishes by hand if the dishes are not prerinsed and if only full loads are washed in the dishwasher.

Without counting lawn watering, typical percentages of water use for a family of four are:

- Toilet flushing—40%
- Bath and shower—32%
- Laundry—14%
- Dishwashing—6%
- Cooking and drinking—5%
- Bathroom sink—3%

In the United States, the National Energy Act of 1992 requires low-volume toilets in new construction or as replacement in existing homes after January 1, 1994. Businesses were to have complied by 1997. Ultra-low-flow (ULF) toilets are available that use only 1.5 to 1.6 gallons (6 liters) for each flush. Low volume toilets are not required in Canada, but water-efficiency plans are in place in many provinces. Plans for new construction are reviewed by appropriate officials in the government.

Sometimes when water conservation is practiced throughout a community, as noted in Question 129, water rates must be increased to provide the money needed by the water utility.

(See Questions 49 and 115 for related information.)

131. Some people say I should put a brick in my toilet tank to save water. How does that save water and is it a good idea?

Toilet flushing uses a lot of water, and putting something in the toilet tank that takes up space means that less water will be used each time the tank refills after a flush, but putting a brick in your toilet tank is not a good idea. A brick tends to crumble and might damage the toilet's flushing mechanism. Instead, use a glass jar, a plastic bag, or a jug filled with water. Because some toilets require a certain volume of water to work right, be sure to test the toilet to make sure it's still flushing well after any changes. *NOTE: Never use your toilet as a trash can. Using several gallons of water to get rid of a tissue or a cigarette is very wasteful.*

Also remember that toilet tanks can leak. To check, put a few drops of food coloring in the tank, wait about 15 minutes, and look in the bowl. If the food coloring shows up there, the tank is leaking and should be fixed. Toilets should be checked for leaks every year. Many large utilities give away conservation kits with food coloring tablets, flow restrictors, and plastic bags to fill with water and put in toilet tanks.

132. I heard that it's a good idea to control the flow of water from my showerhead. How do I measure how fast my shower is using water?

You need two things: a bucket and a watch that can time seconds. The bucket needs to have a 1-gallon (3.8-liter) mark on it. If it doesn't, add a gallon of water and mark the level.

Set the shower flow just as you would when showering. Put the empty bucket under the showerhead to catch all the water and hold it there for 24 seconds (having someone else hold the watch probably will help make this easier). The bucket will weigh 8 to 10 pounds (3.6 to 4.5 kilograms) (10 to 12 pounds in Canada [4.5 to 5.4 kilograms]) after the 24 seconds, so be prepared.

If the water is near the 1-gallon mark, your showerhead is flowing at the recommended amount. If the level is way over the 1-gallon mark, you should consider a new low-flow showerhead (flow restrictors often produce a weak spray) to conserve water. The National Energy Act of 1992 requires low-flow showerheads (less than 2.5 gallons each minute) in any new construction and replacement after January 1, 1994.

Conservation kits are available from a variety of sources. For example, a 1.5-gallon-each-minute showerhead is available by calling (800) 762-7325 and asking for item 46-109, $19.95. This conservation kit contains four other conservation items.

(See Question 129 for related information.)

133. Which uses more water, a tub bath or a shower?

That depends on many factors: how big your tub is, how long you shower, how fast the water comes out of your showerhead, whether or not you turn off the water while soaping, and so forth. Answer this question yourself by closing the drain when you shower and see if you get a tub full of water. Don't try this in a shower stall.

134. I leave the water running while I brush my teeth. Does this waste much water?

You bet! Leaving the water running is a bad habit; about 4 to 6 gallons (20 to 25 liters) of water go down the drain needlessly every time you brush. Turning off the water when you are not using it will save water and save you money.

Another way many people unthinkingly waste water is while they are waiting for the hot water to come to a shower, tub, or sink. Catching this water to use for plant watering is a good conservation tip.

(See Question 90 for related information.)

135. I use a lot of water in the kitchen. How can I conserve there?

Here are several tips:
- Scrape dishes without using water and don't rinse them before putting in the dishwasher.
- Clean vegetables in a pan of water rather than under running tap water, then use that water to give your plants a drink.
- Use the garbage disposal sparingly.
- Run the dishwasher only when it is full.

136. Why are there aerators on home water faucets?

When mixed with water, tiny air bubbles from the aerator prevent the water from splashing too much. Because the water flow is less, often half the regular flow, aerators also help conserve water.

137. Many water quality problems in the home— lead, red water, sand in the system, and so forth—are cured by flushing the system. Isn't that a waste of water?

Yes, but you can avoid losing this water by catching it in a container and using it for plant and garden watering. Even if you don't do this, strictly speaking the flush water is not wasted. A true waste of water is a use that gives no benefit, like leaving the water running while you brush your teeth, setting your lawn sprinkler so the water lands on your driveway or street, or flushing the toilet to get rid of a tissue. Flush water does provide a benefit if it keeps lead or rust out of your water or brings hot water to your tub. Try to use your flush water, but if you can't, don't feel too bad. This water has served a useful purpose.

138. My water faucet drips. Should I bother to fix it?

Yes. Drips waste a precious product, and this waste should be stopped, even though the dripping water may not register on your water meter. To find out how much water you're wasting, put an 8-ounce (236-milliliter) measuring cup (or anything that will let you measure 8 ounces) under the drip and find out how many minutes it takes to fill it up. Divide the filling time into 90 (90 ÷ minutes to fill) to get the gallons of water wasted each day.

As an example, if you have a faucet that dripped 60 times a minute (once each second) this adds up to over 3 gallons (12 liters) each day or 1,225 gallons (4,630 liters) each year, enough to fill more than twenty-two 55-gallon (210-liter) drums, just from one dripping faucet. This leak would fill the 8-ounce (236-milliliter) measuring cup in less than 30 minutes.

139. How should I water my lawn to avoid wasting water?

Water your lawn for long periods a couple of times each week, rather than every day. This allows deep penetration of the water. An inch a week is a good rule of thumb, but this varies for different grasses and different parts of the country. Check with your local garden store. If you want to find out exactly how long to water, put some large cans or jars (peanut butter jars will work) around your lawn and see how long you have to run your sprinkler to fill the jars with the right amount of water. During times of drought, your water supplier may set allowable times for lawn watering. When this happens, be sure to follow the rules as long as they are in force.

Water early in the morning to avoid excessive evaporation; it is usually less windy then, too, and the water pressure is usually higher. Except in the arid West, night watering may promote lawn disease. Use a sprinkler that makes large drops, because small drops evaporate faster. Some areas of the country permit the use of wastewater treated beyond typical treatment to be used for watering. Watering your lawn with a hand-held hose is a waste of both your time and your water, although it might be okay for a small garden.

Try to avoid watering paved areas and don't use your hose to wash sidewalks or driveways. Both of these practices waste a lot of water.

Plant watering can be reduced by selecting **xeriscape** (which means low-water-demanding)  plants or native plants, which provide an attractive landscape without high water use. Two good references are: (1) *Xeriscape Handbook* by Gayle Weinstein; and (2) *Xeriscape Plant Guide*. Both books were published by Fulcrum Publishing and are available from AWWA by calling (800) 926-7337.

Remember, pets need water in the summer just as your lawn and garden do. If you keep a pet outdoors, provide plenty of water in a shady area. Secure the water bowl so that it will not spill. Add ice cubes from time to time to keep the water cool.

140. I have a private well. I don't need to conserve water, do I?

Oh yes, you do! The underground source you are using is limited, so every drop counts. Please save as much as possible so that water will be left for others.

141. Why do we still have a water shortage when it's been raining at my house?

Your drinking water may come from far away, so even if it's raining at your house, it may not be raining where the water supplier collects its water. If this happens, the rain in your area doesn't help the water shortage, but it usually does lower the demand for water while it is raining because people stop watering lawns and gardens and washing their cars.

(See Questions 148 and 149 for related information.)

142. During times of water shortage, shouldn't decorative fountains be turned off?

In most cases, fountain water is recirculated (used over and over) and is not wasted. If water losses from evaporation are high, however, fountains should be turned off.

143. During times of water shortage, does not serving water to restaurant customers really help?

Skipping water in restaurants serves as a good reminder to everyone about the importance of saving water, but the actual volume of water saved is small. The water that is used to wash the water glasses is also saved, and this is usually more than the drinking water that would normally be served by glass to customers (two glasses of water for each glass washed). By comparison, each flush of a toilet uses the equivalent of about 80 glasses of drinking water.

Health experts emphasize the importance of drinking at least six to eight glasses of water each day (remember you drink water in many forms—see Question 21), so if you want water to drink, ask for it. Skip it if you are just going to let the glass sit on the table while you drink something else.

A good reference on conservation is: *A Consumer's Guide to Water Conservation: Dozens of Ways to Save Water, the Environment, and a Lot of Money,* published by AWWA. Call (800) 926-7337 and ask for item No. 10063, $4.50 member, $6.75 nonmember.

(See Questions 64, 121, 129, and 154 for more comments about conservation.)

144. Questions 148 and 149 imply that the amount of water on the globe isn't changing, so why should I conserve?

You're right about the amount of water being constant, but conserving is still important. Here's an example. Suppose you're in a growing community. As the population increases, so does the demand for water. This means that every so often, the water supplier must spend some money to find another source of water and in some areas additional sources are hard to find. If people conserved, the water demand would not grow as fast as the population and the need to look for more water would be delayed. This permits the municipality to defer expenditures and to use the money for something else in the meantime. In addition, not all of the water taken as drinking water gets right back into the source. Thus, if a community is conserving water so that less is needed, more water will be left as an aquatic habitat for creatures that live in water.

(See Question 154 for related information.)

5
Sources

Water is more precious than gold and more explosive than dynamite.

—E. K. McQuery

GENERAL

145. Where does my drinking water come from?

There are two major sources of drinking water: surface water and groundwater. Surface water comes from lakes, reservoirs, and rivers. Groundwater comes from wells that the water supplier drills into aquifers. An aquifer is an underground geologic formation through which water flows slowly. Some wells are shallow—50 to 100 feet (15 to 30 meters) deep; others are deep—1,500 to 2,000 feet (450 to 600 meters).

Springs are another source of water. Springs begin underground as groundwater. When the water is pushed to the surface and flows out of the ground naturally, it becomes a spring. The water then may flow over the surface of the ground as surface water.

Most large cities in the United States—New York, Los Angeles, Chicago, Philadelphia, Boston, St. Louis, New Orleans, Seattle, Denver, and part of Houston—to name a few, use surface water. The larger cities in Canada, including Montreal, Toronto, Edmonton, and Vancouver, also use surface water.

Most small towns use groundwater. About 80 million people in the United States use groundwater supplied through individual wells and municipal groundwater systems. About 2 million people in Canada are supplied by municipal groundwater systems.

Some water suppliers buy treated water from others (wholesalers) and then provide water to their customers, often without further treatment.

Your local water utility can tell you the specific source of your drinking water. Also, in the newly required annual "Consumer Confidence Report," the source of your water will be listed.

146. I have heard the term "mining groundwater." Is that anything like mining coal?

Yes, mining water is similar to mining coal. As coal in mined, less and less remains in the ground. Although underground water (groundwater) is replenished by rainfall soaking into the ground, this is a slow process. Often groundwater is pumped out of the ground faster than it can be restored; thus, the groundwater is being "mined." When this happens, the underground level of the groundwater falls, and wells have to be drilled deeper to reach it.

(See Question 145 for related information.)

147. How much drinking water is produced in the entire United States each day, and how does that compare with the water used for industrial purposes and irrigation of crops?

Almost 40 billion gallons of tap water is produced each day for domestic use (homes, restaurants, hotels, small businesses, and so forth) in the United States, 60 percent from surface water and 40 percent from groundwater. Daily irrigation use is much larger, but the volume depends on the location and the time of year. It takes about 50 glasses of water just to grow enough oranges to produce one glass of orange juice, for example. One estimate puts the total amount used for irrigation at 141 billion gallons a day, 66 percent from surface water and 34 percent from groundwater. Of course, irrigation water is not treated as tap water is. Finally, industrial use is about 160 billion gallons per day.

148. Are we running out of water?

Globally, we have sufficient fresh water to satisfy the need for drinking water, but frequently it is not located where the high-use areas are. Thus, localized water shortages occur. Furthermore, *droughts* (below-normal rainfall), often lasting several years, worsen water shortages in some areas.

149. How does nature recycle water?

The *water cycle* keeps the amount of total water on the globe constant. Water from oceans, lakes, rivers, ponds, puddles, and other water surfaces evaporates to become clouds. The clouds make rain, snow, or sleet that falls to earth to make rivers and streams, some of which seeps into the ground to form groundwater. All of this water flows to the ocean to start the cycle over again.

Before returning to the ocean, some of this water is taken for drinking water and then is discharged as wastewater. The cycle is never-ending.

150.

When I see pictures of the earth from outer space, It looks like it's mostly covered with water. Is that right, and how much of this water is drinkable?

You're right. Between 70 and 80 percent of the earth is covered with water, but less than 1 percent of this visible (surface) water can be made into drinking water with conventional treatment (See Question 170). Remember, however, that great quantities of water are available underground that cannot be seen from space. This water is called groundwater. In the United States alone there is an estimated 253 billion gallons of fresh water used for all purposes every day from surface water and 88 billion gallons used from groundwater. So you can see how important this "invisible" groundwater is.

Of all the water in the world, 97 percent is in the oceans; 2 percent is in ice caps and glaciers; 0.3 percent is deep underground (unusable); 0.3 percent is underground relatively near the surface (about half of this has a low enough salt content to be used as a source of drinking water); 0.3 percent is in the atmosphere; and 0.1 percent is in rivers, lakes, and reservoirs.

151. Many areas near the ocean do not have large supplies of fresh water. Why can't ocean water be treated to make drinking water?

Ocean water can be treated, but the process is expensive. On the other hand, so is the economic cost of not having enough water. The cost of converting salt water to drinking water has been estimated at $5 to $7 for each 1,000 gallons instead of the $0.30 to $0.50 for each 1,000 gallons for the treatment described in Question 170. Of course, the cost of ocean water treatment must be compared to the "total" cost of providing water from another source, building dams and pipelines, and so forth. Ocean water contains so much salt that at least 99.2 percent of the salt would have to be removed to avoid a salty taste in drinking water.

152. Why can't we just drink ocean water?

Humans are made mostly of water. Inside our bodies are millions of tiny balloon-like cells. If those cells were put into salty ocean water, the water inside the cells would pass right through the cell walls into the ocean water. This process is called **osmosis**. The cells would then shrink and die. This would happen in your body if you drank ocean water, and you would die.

153. I've heard about towing icebergs to areas that are short of water as a source for drinking water. Would that really work?

It's never been done commercially as it would be so expensive, but it would work. Even though icebergs are floating in salt water, the ice has no salt in it—it's compressed snow. If you melted an iceberg you would get drinkable fresh water after you killed any germs.

154. Can wastewater be treated to make it into drinking water?

This is not done at the present time, although a recent test showed that good quality drinking water could be made from wastewater. But so far, water shortages have not been so severe that this measure has been needed. Of course, nature reuses water through the water cycle.

To save water in some areas of the United States, wastewater treated beyond conventional treatment is used to irrigate golf courses and landscaped public areas, provide cooling water for industry, and occasionally for flushing toilets, although this practice is not without controversy. This irrigation practice is not promoted in Canada.

QUALITY

155. Which is more polluted, groundwater or surface water?

It depends on what you call pollution. Because surface water can be contaminated by municipal sewage, industrial discharges, transportation accidents, and rainfall runoff, it contains many pollutants but not much of any one chemical.

Groundwater, on the other hand, may contain pollutants such as arsenic, nitrates, radioactive materials, and high (compared with surface water) amounts of a few organic chemicals such as cleaning fluid or gasoline. Therefore, both may be polluted but in different ways. Another difference is that the degree of pollution may change rapidly in surface waters, while pollution levels change very slowly in groundwater. Your water supplier can tell you what contaminants it has found in its source water, but it tests the quality of its treated water more than its source water.

(See Question 57 for related information.)

156. In towns and cities, what is the major cause of pollution of drinking water sources?

The major source of pollution is rainwater that flows into street catch basins (called *urban runoff* or *stormwater runoff*). While this rainwater alone is not necessarily harmful, it frequently carries untreated waste products from our streets and yards directly into rivers, streams, and lakes (drinking water sources).

157. Why isn't urban runoff usually treated before being discharged into drinking water sources?

Because runoff from rainfall occurs only when it rains, it doesn't make sense to build special treatment processes for those times when it isn't raining. As an analogy, having extra servers at a restaurant all day to avoid any delays at lunch time is not practical, as they would be idle at other times of the day. Chicago has one possible solution—capturing and holding the runoff during storms and then pumping it slowly and steadily to the wastewater plant for treatment prior to discharge. Other cities use ponds to hold the runoff for a while so that its quality can improve by settling out some contaminants before discharge. Because runoff is an important source of pollution, legislation that deals with stormwater runoff is changing the way it must be treated, in spite of the difficulties.

158. Does acid rain affect water supplies?

Some air pollutants do dissolve in raindrops and make them *acidic*, like lemon juice, but the effect of this acid rain on water supplies is small. Certain lakes in the northeastern United States and in certain regions of Canada seem to be a little more acidic than they were years ago, but the changes are minor and current rules require water suppliers to remove any acidity before the water goes to the consumer. Groundwater usually is not affected because the alkaline materials in soils (like the chemicals in antacid tablets) react with the acid rain to neutralize it before it reaches the groundwater. An exception to this rule occurs in northern Ontario, Canada, where the groundwater is near the surface. In that location, there is not enough soil to neutralize the acid rain before it reaches the groundwater. Even though the groundwater in this area is a little acid, most of the utilities here use surface water, which is fine. The few municipalities that do use groundwater add chemicals to react with the acid and destroy it to avoid acid (corrosive) water. Most of the groundwater use in the area is by individuals with private wells, and most of them do nothing to treat the water.

(See Questions 61, 63, 86, 119, and 175 for related information.)

159. I read about the problem of oil spills. Do they pollute drinking water sources?

Although oil spilled in the oceans is bad for the environment, it is not a danger to drinking water sources. However, ship and barge accidents can contaminate surface water sources (rivers and lakes). Many highways and railroad tracks pass over drinking water sources, creating a potential for contamination if a truck or freight-train accident occurs. A motor vehicle accident or poor disposal of oil from your car can cause oil pollution. Drinking water contaminated with even a little bit of oil has such a bad taste that most people regard it as undrinkable.

Although groundwater sources are not directly affected by most of these types of accidents, unless they occur in the *recharge area* (where water seeps downward to add water to the underground aquifer), major oil spills in these areas can cause nearly irreversible damage to groundwater.

160. I live downstream from a nuclear power plant. Should I worry about radioactivity in my drinking water?

No, for two reasons. First, the nuclear power plant operates under strict guidelines from the Nuclear Regulatory Commission to prevent dangerous radioactivity from getting into the water, and second, the radioactive content of drinking water is regulated by the US Environmental Protection Agency to prevent excessive amounts from being in tap water.

(See Questions 75–77 for related information.)

161. How can I help prevent pollution of drinking water sources?

Properly dispose of the chemicals you use in your home. Every chemical you buy has the potential of polluting the environment if disposed of improperly. Try to buy environmentally acceptable alternative products and, to minimize waste, buy only what you can use. Many larger cities have a hazardous waste disposal department. Check with it if you have disposal questions. If you change the oil in your car yourself, find out from your city, state, or provincial environmental agency how to properly dispose of the used oil.

Remember, if your home is served by a sewage system, your drain is an entrance to your wastewater disposal system and eventually to a drinking water source. Discharges from septic tank drain fields may pollute groundwater. Treat your wastewater system with respect.

(See Questions 62, 117, and 118 for related information.)

162. Why does my water sometimes have sand in it?

Routine cleaning of pipes that carry drinking water can stir up material that has settled to the bottom of the pipes. This can give your water a temporary sandy appearance. Some wells have specially designed screens (like window screens) to hold back the sand, and a break in this screen is another possible cause of sandy water. Small quantities of sand may pass through a well screen, even if it is not broken. The best way to solve this problem is to verify with your water supplier that there is no break in the system, and if there is none, flush your home pipes by running water for a while through your largest faucet, probably in the bathtub. Saving this water for other uses is a good conservation measure. Some particles that look like sand may actually result from the corrosion of galvanized pipes in your household plumbing.

(See Questions 64, 86, and 121 for related information.)

6

Suppliers

Men work on earth at many things;
Some till the soil, a few are kings;
But the noblest job beneath the sun
Is making Running Water *run.*

—John L. Ford, *Water and Wastewater Engineering*

GENERAL

163. How many drinking water suppliers are there in the United States and Canada?

There are approximately 200,000 community and noncommunity drinking water suppliers in the United States. The US Safe Drinking Water Act (SDWA) defines a *community* water supplier as a utility that provides drinking water to 25 year-round residents or more all the time. There are just under 60,000 community water suppliers in the United States. Most are small; about 63 percent are in towns with 500 people or fewer. About 250 community suppliers serve populations of 100,000 or more—that is, about 110 million Americans in all. In addition, there are about 140,000 *noncommunity* drinking water suppliers—motels, remote restaurants, and similar establishments that serve the traveling public and water suppliers for schools, factories, and so forth. While they too may serve many people, they don't serve them on an ongoing basis. All of these 200,000 drinking water suppliers must meet some or all of the federal drinking water regulations. Because you don't stay in one place for very long when you are traveling, noncommunity supplies don't have to meet as many federal regulations as does your home supply where you drink the water almost all of the time.

In the United States, about 13 million people (5 percent of the population) are supplied water by their own private wells. These wells are not federally regulated under the SDWA.

In Canada, a municipal supply is defined as one serving five households or more. There are about 5,000 municipal supplies, with about 60 percent serving small communities. In addition, there are about 50,000 private communal supplies that serve the traveling public and small permanent facilities. About 4 million Canadians obtain water from their own private wells.

(See Question 184 for related information.)

164. Are all drinking water utilities owned by cities and towns?

No. In the United States, over half of all drinking water suppliers are privately owned, just like any other company. The amount these companies charge their customers for drinking water is controlled by state public utilities commissions. Many of these companies are listed on the New York Stock Exchange or other exchanges, and the public may purchase shares. In Canada, almost all of the municipal supplies are publicly owned, although recently the federal government has established the council for public–private partnerships so the situation may change in the future.

(See Question 126 for related information.)

165. I've read all about the problems with my drinking water. Why isn't my water supplier doing something about it?

There are several possible answers:
- The utility may be working on the problem but not informing consumers; construction of new facilities takes a long time.

- The utility may want to do something about it but doesn't have the financial resources.
- Because drinking water suppliers are dedicated to providing the best quality drinking water at the least cost to the customer, they may be waiting until all federal regulations are in place before taking action to avoid wasting money on obsolete or unnecessary procedures.

Contact your utility to find out about your specific situation.

(See Chapter 8 on federal regulations for related information.)

166. My drinking water is not acceptable. To whom do I complain?

Contact your local drinking water supplier and discuss your problem. Your supplier will probably send a representative to your home to help you or to explain what is causing the problem and what the supplier intends to do about it.

167. How does a water supplier prepare for a possible natural disaster, such as a hurricane?

If the water supplier has some warning, the supplier will make sure that all necessary chemicals and supplies are available, that the emergency power systems are well supplied with fuel, and that all water storage tanks and basins are filled. Water suppliers have contingency plans to make sure that a continuous supply of drinking water will be available under almost any circumstance.

Many water plants that are built in a floodplain are diked to avoid being flooded. The extraordinary floods of 1993 in the midwestern United States and in 1994 in southern Georgia were so high, however, that the dikes could not protect all of the water treatment plants. After an event like that, it takes a long time to deliver *safe* drinking water to the community again because even after the plant is back in operation, killing *all* of the germs in *all* of the pipes underground is very difficult. Another problem with floods is the contamination of wells (both public and private) that are covered by flood water. The water is not safe to drink until extensive testing proves that all is okay again. You will be informed by the water supplier when you can start drinking the water again.

168. What should I do to help my drinking water supplier improve my drinking water quality?

You can do several things:

- Contact your local drinking water supplier and ask what the major needs are and how you can help.
- Support governmental efforts, such as increases in water rates from your water supplier to improve service, local taxation and bond issues to improve drinking water quality, and similar efforts.
- Write to your state, provincial, or federal legislators urging them to increase funding for support of legislation protecting drinking water quality.
- Discuss these issues with your friends and associates and get their support as well.

169. Can my drinking water be sabotaged?

Water utilities have always taken seriously the task of providing safe, adequate, and secure water supply to consumers. Most utilities are aware of potential sabotage events and have taken measures, including monitoring and testing, to keep the water supply safe and their facilities secure. In the unlikely event that a water supply is intentionally disrupted or contaminated, utilities also have emergency plans that address such issues as redundancy of operations, alternative water sources, and public notification. If there was an emergency situation, such as a contamination event, you would be notified by your utility to either boil water or cease using it, and would be supplied with either bottled water or water brought in on tankers called water buffaloes, until the event was under control. Because the volumes of water treated are generally large, the quantity of a contaminant that would have to be added to pollute a water supply is so great that sabotage is almost impossible.

As an example, in August 1994, a Colombian national was arrested in Germany carrying 14.4 ounces (408 grams) of 87 percent **plutonium**. Some newspapers suggested that this could be used to poison the water supply in Munich, Germany. Let's do a little figuring. Munich stores only one day's supply (very little compared to most systems). So 35.2 ounces (1,000 grams—2.4 times the amount seized) would dilute down to 0.003 mg/L (see Question 188 for the definition of mg/L). Drinking 2.1 quarts (2 liters) of this water would produce a radiation exposure of about 1 percent of what we get *naturally* in a year—not too dangerous.

TREATMENT

170. How is my drinking water treated?

Currently in the United States about 50 percent of groundwater supplies are distributed untreated. In the near future, however, the US government will adopt regulations that are expected to force many more groundwater systems to add a disinfectant, which is done to kill germs.

Most surface waters first are treated with chemicals that combine with dirt particles. The dirt then can be settled and filtered out of the clean water, making it clear. Filtering is important because, besides making the water clear, it removes some germs that are difficult to kill. Finally, the water is disinfected to kill the remaining germs. Thus, tap water is a manufactured product, starting with a raw material (surface water) and processing it to turn it into a finished product, your drinking water. Groundwater is usually clear when it is pumped out of the ground; thus, it can be disinfected without prior treatment. Occasionally, additional treatment is needed to solve special problems. The treatment described above does not remove dissolved toxic chemicals, so it is used only when they are not present, the most common case.

In 1994, a report entitled "Victorian Water Treatment Enters the 21st Century" was released to the press. One point made in this report was that "obsolete" treatment techniques were being used in this country. Although it is true that the treatment methods described above are the same ones that were in use in the early 1900s, the improvements that have been made in them over the years have dramatically improved their performance, yielding a much higher quality drinking water. Additional treatment beyond what is now "conventional" is expected in the future, moreover, as water suppliers take strides to make drinking water even safer than it is today.

Many different chemicals are used by your supplier during the water treatment process. Some of these chemicals are removed along with the pollutants; others remain in the water. The ones left, however, are either nontoxic or their use is strictly limited.

The same treatment situation exists in Canada, but some Canadian provinces require disinfection as the minimum treatment.

(See Questions 26, 58, 73, and 186 for related information.)

171. Is it true that some major cities do not filter their drinking water?

According to US Environmental Protection Agency regulations, since June 29, 1993, all water suppliers using surface water have been required to use a filtration step in their treatment. The regulation does, however, allow water suppliers not to filter their water, if they can show, by passing a complex set of tests, that their drinking water is safe without filtration. Filtration has been installed in many locations recently and is being considered in many others. If you are interested in your local situation, talk to your water supplier.

In Canada, a similar situation exists. General policy requires filtration of surface water, but a local supplier that can demonstrate it meets stringent testing, reporting, and quality criteria need not filter. As in the United States, many localities are assessing the need for filtration.

(See Questions 26 and 145, and Chapter 8 for related information.)

172. What is the best disinfectant for my water supplier to use?

As noted in the answer to Question 72, there is no one answer. Each of the common disinfectants has advantages and disadvantages. Every water supplier must consider all of these advantages and disadvantages before choosing a disinfectant for a particular water supply. If you want to know what disinfectant is used in your system, contact your water supplier.

7

Distribution

Sub-human primates, like other animals, drink where the water flows.

Only man carries it to where he lives.

—C. S. Coon, *The History of Man*

173. Are the pipes that carry drinking water from the treatment plant to my home clean?

A well-run water utility will have an ongoing program of flushing and cleaning the distribution pipes to ensure that these pipes are clean. Otherwise, rust and microbes would cling to the pipe walls. Flushing is done by opening fire hydrants and letting the water rush out.

Another way to clean pipes is by forcing a tight-fitting plastic sponge through the pipe using water pressure. The sponge scrapes the pipe walls clean. The dirt then is flushed out a fire hydrant. Similar devices with other designs are also available.

Keeping water pipes clean is a big job, as there are about 1 million miles (1.7 million kilometers) of pipes in the United States and Canada.

174.

I have seen work crews cleaning water mains and the water they flush out looks terrible. How can the water be safe if the pipes are so dirty?

Almost all water pipes have a thin film of rust and harmless microbes on the inside. Experience has shown that this thin film causes no problems. Buildup of this material may, however, cause problems such as clogging of fixtures, causing the tap water to look bad, or using up the disinfectant in the water as it passes through the pipes. Water suppliers have a regular program of flushing and cleaning their distribution pipes. When they remove all of this material from the walls of several miles of a pipe and it comes out a fire hydrant all at once, it looks worse than it really is. If you watch the workers do this, you will notice that the water clears up rather quickly.

(See Question 26 for related information.)

175. What are asbestos–cement pipes that carry water under the streets? Are they safe?

Asbestos–cement pipes are made from cement with asbestos fibers added to make the pipes strong. Most drinking water passes through these pipes without becoming contaminated with asbestos fibers, and so they are safe. Although asbestos has been banned for many uses by the US government because of the health risks caused by breathing asbestos fibers, asbestos–cement pipes are exempted. However, they are not popular in new construction. Cement-lined iron pipe is the most common type these days.

A few types of drinking water—water that is slightly acid or contains few minerals—tend to dissolve the cement in asbestos–cement pipes, softening the pipe and releasing the asbestos fibers into the drinking water. Asbestos–cement pipes should not be used to carry these types of drinking water. Because of the US Environmental Protection Agency's (USEPA) lead and copper regulation, water quality must be adjusted to make sure that the water is not corrosive, thus helping prevent fiber release. Remember, if the pipe stays solid and is not softened by the water flowing through it, the fibers are so tightly bound in the pipe material that they will not be released and the pipe is safe.

Additionally, the USEPA limits the number of asbestos fibers permitted in drinking water. This is important because studies have shown that ultrasonic humidifiers using water that contains asbestos fibers may put fibers into the air where they might be dangerous because they are inhaled. Asbestos fibers are much more dangerous when inhaled than when they are swallowed.

Asbestos fibers in drinking water are not regulated in Canada, and except for avoiding slightly acid waters, asbestos–cement pipe can be used.

To learn about water conditions in your area and to find out if asbestos–cement pipes are used, call your water supplier.

(See Questions 61, 63, 101 and Chapter 8 for related information.)

176. Why are fire hydrants sometimes called fire plugs?

Long ago, drinking water was distributed through towns in wooden pipes. When water was needed to fight a fire, a hole was drilled in the wooden pipe. When the fire was over, a wooden plug was used to close the hole. These *fire plugs* were then marked for possible future use.

177. We don't use much water for drinking. Why does all the rest need to be treated so extensively? This seems unnecessarily expensive.

Actually, the cost of water treatment is a small part of your water bill (only about $0.75 to $1.50 each month for each person living in your house), but that is not the real answer. A new town could be developed in which two pipes come into your home from the street (a dual distribution system), one small pipe for drinking water and a larger pipe for all of the other uses for water (toilet flushing, lawn watering, and so forth). To take out all of the pipes and install such a system in existing cities would cost far too much. Some smaller communities are modifying their systems, however.

In any case, the water provided for purposes other than drinking would still need to be free of germs and any chemicals that would stain laundry or form deposits on pipes. Thus, even this water would require some treatment.

(See Question 126 for related information.)

178. How does a water company detect a major leak in the distribution piping system?

A major leak can be detected by:

- Visual detection (water on the ground) by water company employees who work in the field.
- A loss in pressure that can be detected by the water company and customers.
- Reports by public-minded citizens.

Once a leak is suspected, its precise location is determined by water utility personnel. Sensitive listening devices are used to detect the sound of the leaking water underground.

Stopping leaks is important to a water supplier because leaks waste water, adding cost to both the water supplier and you. The water supplier doesn't get paid for the water that is lost to leaks, but may pass its cost along to the customers. The national average for water lost from leaks is 15 percent, although most suppliers try to keep such losses to around 10 percent. Gas companies only lose about 5 percent of their product.

Any leakage that occurs within the boundary of your property after the water meter is your responsibility and must be repaired at your own expense. Prompt repair is to your benefit, because as long as the pipe is leaking, your water bill will be higher.

179. If leaks are such a waste of water, why doesn't the water utility just fix them?

Your water supplier is certainly trying to minimize leaks because they do waste water. Leaks may result from old rusting pipes or from ground movement that causes pipes to break or

their joints to crack. In the complex network of pipes necessary to supply water throughout a community, it is almost inevitable that at any one time some leakage will be occurring. Many water departments have special teams to measure leakage, locate and repair leaks, find weak points in the pipe network, and replace rusting pipes. Leakage also can be controlled but not eliminated by the supplier's avoiding unnecessarily high water pressure.

180. Fixing a broken water pipe looks like a dirty job. How is the inside of the pipe cleaned afterward?

After work is done, the pipe is filled with water containing a large amount of chlorine. Holding this water in the pipe for a time kills all the germs.

This is not the end of the story, however. The next problem is how to dispose of all this water that contains so much chlorine. State, provincial, and federal regulations control its disposal. A chemical must be added to react with the chlorine and destroy it before the water can be flushed out of the pipe and discharged, or the highly chlorinated water must be discharged to an area where it will not have an adverse impact on the environment.

(See Question 26 for related information.)

181. What are the causes of low water pressure and should low water pressure concern me?

Temporary low pressure can be caused by heavy water use in your area—lawn watering, a water main break, fighting a nearby fire, and so on. Permanent low pressure could be caused by the location of your home—on a hill or far from the pumping plant—or your home may be served by pipes that are too small, or the pipes in your home have a lot of scale (see Question 102) in them, leaving little room for the water to flow. This is more common in older homes.

Low pressure is more than just a nuisance. The water system depends on pressure to keep out any contamination. If the pressure drops, the possibility of pollution entering the drinking water increases. One of the causes of poor quality water in some developing countries is low pressure in the distribution system that allows contamination to enter the pipes. You should report any permanent drop in water pressure to your water company.

Many areas have minimum standards for pressure. For example, 20 pounds per square inch (psi) (140 kilopascals) when water use is at a maximum is a common standard (car tires often use 30 to 32 psi of air [about 210 kilopascals]). Most systems have pressures three to four times the minimum.

You can tell you may have low pressure if flows from your faucets at home are much lower than elsewhere in your area—at work, in a restaurant wash room, or in a friend's home elsewhere in the city, for example. The only way to cure general low pressure is to have the supplier change the system, adding more pumps or bigger lines, but because low pressure is a possible health hazard, keeping the pressure up is important. If the problem is in your home (more of a nuisance than a potential health hazard), discuss your options with a reputable plumber.

You may be surprised to learn that you can also have too much pressure. Some homes need pressure regulators to avoid damaging household plumbing from very high water pressures.

182. What are cross-connections and why are they a problem?

A cross-connection is a connection between a drinking water pipe and a polluted source. Here's a common example. You're going to spray weed killer on your lawn. You hook up your hose to the sprayer that contains the weed killer. If the water pressure drops at the same time you turn on the hose, the chemical in the sprayer may be sucked back into the drinking water pipes through the hose. This would seriously pollute the drinking water system. This problem can be prevented by using an attachment on your hose called a backflow-prevention device. This is a way for consumers to help protect their water system. In some areas, this is mandatory for new construction.

Most water suppliers have cross-connection control programs, particularly in major cities. Their distribution systems are so complex that tracking down cross-connections is a never-ending job. Removing cross-connections is vital, however, if drinking water quality is to be protected.

183. Why is some drinking water stored in large tanks high above the ground?

Two reasons. First, this type of storage ensures that water pressure and water volume are sufficient to fight fires, even if the electricity that runs the water pumps is off. The second reason is to provide an extra source of drinking water during the day when water use is high. The tanks are refilled at night when drinking water use is low. Water suppliers must be very careful of these storage tanks because water may stay in them a long time. Thus, there is a potential for a decline in water quality. Regular inspection and sampling will prevent problems.

8
Regulations and Testing

It was Emperor Nero's invention to boil water, and then enclose it in glass vessels and cool it in the snow. Indeed, it is generally admitted all water is more wholesome boiled.

—John Bostock and H. T. Ripley,
The Natural History of Pliny

FEDERAL REGULATIONS

184. What federal legislation protects the quality of drinking water?

In the United States, the Safe Drinking Water Act (administered by the US Environmental Protection Agency), first passed in 1974 and expanded and strengthened in 1986 and 1996, protects the quality of drinking water.

In Canada, drinking water is a shared federal–provincial responsibility. In general, provincial governments are responsible for an adequate, safe supply, whereas the Federal

Department of National Health and Welfare develops water quality guidelines through a joint federal–provincial mechanism. These guidelines, however, are not legally enforceable unless adopted by a provincial agency. Two Canadian provinces, Alberta and Quebec, have adopted the federal guidelines as regulations.

185. Who enforces the federal standards?

In the United States, in general, your state health department has responsibility for enforcing the federal standards, although under certain circumstances another governmental agency may have that duty. Your water supplier can tell you who is responsible in your state. The states must adopt drinking water quality standards that are at least as strict as those of the federal government. Each state then evaluates its own water supplies and ranks them as *in compliance* or *out of compliance.* Of course, those out of compliance must be improved. A few states post signs at the city limits stating if the drinking water there is "in compliance."

In Canada, the appropriate provincial health and environmental agency has responsibility for the safety of drinking water and the application of the appropriate drinking water quality guidelines.

186. Is water that meets federal drinking water standards absolutely safe?

Safety is relative not absolute. For example, an aspirin or two may help a headache, but if you took a whole bottle at once, you'd probably die. So, is aspirin safe? When setting drinking water standards, federal regulatory agencies use the concept of *reasonable risk*, not risk-free. Risk-free water would cost too much. So the answer to the question is, no, drinking water isn't *absolutely* safe. But the likelihood of getting sick from drinking water that meets the federal standards is very small, typically one chance in a million.

One difficulty the US Environmental Protection Agency has when trying to determine reasonable risk relates to the problem called *susceptible population.* Not all people who drink water are the same from a health point of view, that is, some people are more susceptible to getting sick than others. For example, only babies three months old or younger are affected by nitrates in drinking water, so for that contaminant they are the susceptible population: They are susceptible to getting sick from too much nitrate in their drinking water. The standard for nitrate, therefore, was chosen to protect these infants. With other contaminants, identifying the susceptible population is not as easy. Are they the elderly, those undergoing cancer treatment, those in nursing homes, all babies, those who are HIV positive, or others? For each standard, the federal regulatory agencies must balance the risk to all these groups against the cost of treatment and arrive at a standard that will protect as many people as possible and that can be afforded. This is called "the greatest good for the greatest number."

(See Questions 37, 41, and 60 for related information.)

187. How do federal regulatory agencies choose the standard for a chemical in drinking water?

Because rats and mice digest their food in the same way humans do, they are affected by toxic chemicals in the same way humans are. Therefore, scientists at the National Toxicology Program of the federal government feed these animals a chemical in question for a two-year period to determine its effects. From this information and using a safety factor, a drinking water standard based on "reasonable risk" is determined. For most potentially cancer-causing chemicals, reasonable risk is defined as follows: If 1 million people drank water for a period of 70 years with the amount of chemical in it equal to the standard, no more than one additional person would probably get cancer *from the drinking water*—a very small risk.

188. Most federal standards are written like this: "Selenium—0.05 mg/L." What does "mg/L" mean?

The abbreviation *mg/L* stands for milligrams per liter. In metric units, this is the weight of a chemical (selenium in the example) dissolved in one liter of water. One liter is about equal to 1 quart, and 1 ounce is equal to about 28,500 milligrams, so one milligram is a very small amount. About 25 grains of sugar weigh one milligram.

Because 1 liter of water weighs 1 million milligrams, 1 milligram per liter (mg/L) is sometimes written 1 part (weight of chemical) per million parts (weight of water), or 1 ppm (part per million).

One part per million is hard to imagine. Here are some examples: 11.6 days contain 1 million seconds, so 1 second out of 11.6 days is 1 part per million (1 second in a million seconds), or 25 grains of sugar in a quart of water is about 1 part per million, or 1 drop in a 55-gallon drum is about 1 part per million. You can see that the weight of selenium allowed by the federal government in a quart of water is very, very small. The limits set for some other chemicals in drinking water are even smaller. For example, the standard for cleaning fluid, a dangerous chemical, is only 0.002 mg/L (25 times less than selenium). Chemicals with such severe limitations are considered a serious health hazard.

Once a drinking water standard has been chosen for any contaminant, it is called the "maximum contaminant level," abbreviated MCL, for that contaminant. In addition to the MCL, by law, the US Environmental Protection Agency must chose a "maximum contaminant level goal" (MCLG) for any regulated contaminant. This is the level below which there is *no* known or expected risk to health. Contrast this with the "reasonable risk" concept. The MCLG also contains a factor of safety. The US Environmental Protection Agency does not enforce MCLGs.

If you would like a list of the federal regulations, look in Appendix A for the location of the US Environmental Protection Agency office nearest you or for the Safe Drinking Water hot line telephone number.

189. Federal, state, provincial, and local governments will spend a lot of money over the next few years to satisfy the requirements of federal laws about drinking water quality. Will any lives be saved?

Much of the money will be spent to prevent germs like *Cryptosporidium* oocysts from reaching consumers. Because germs in drinking water have caused deaths, improving microbial water quality could save some lives. Money also will be spent to remove cancer-causing chemicals from drinking water, especially disinfection by-products.

(See Questions 15, 31, and 70 for related information.)

TESTING

190. How is my water tested and who tests it?

Federal regulations state that all water suppliers must test the treated water for microbes and chemicals (a list of nearly 100 in the United States) a specified number of times each year. The tests for microbes are done most often; the frequency varies depending on the population served by a water supplier. Federal regulations in the United States also state that these tests must be conducted in federally certified laboratories using federally approved methods, some of which are quite complex. In Canada, most tests are carried out in provincial laboratories. Private wells are frequently tested in connection with the sale of a home.

(See Questions 13, 38, and 57 for related information.)

191. Can water systems be excused from monitoring for some contaminants?

In some cases, yes. Over time, by reviewing test results and keeping a watchful eye on potential problems, public water systems come to understand the likely threats to their water supplies. If a water utility does not have water quality problems, it can apply to the state for permission to test less frequently for certain contaminants. If, after scientific analysis, state regulators believe human or natural activities are unlikely to affect the system's water quality, they may grant the request to avoid unnecessary testing. A waiver from some monitoring requirements in no way reduces the water supplier's responsibility to provide high-quality drinking water.

192. Can I test my water at home?

Not in a meaningful way. Simple kits are available to test for some chemicals like chlorine, calcium, and lead, but a thorough analysis is not possible with these kits. In some cities, such as New York, water suppliers have begun offering free tests for lead in water.

193. I have a private well. Who will test my water?

You can telephone your state, provincial, or local health department. Officials will help arrange to have your water tested, or they will explain how to take a sample for microbes and where you can take the sample for testing.

Larger cities have commercial laboratories that will test drinking water for chemicals, but the tests are more expensive and often difficult to understand. The cost will depend on the type of tests you have done. A test for microbes (coliforms) will probably be less than $10, but testing for chemicals can run into hundreds of dollars.

At the very minimum, you should have the water tested for lead, as well as coliforms, nitrate (particularly if you have young children), radon, and arsenic.

(See Questions 28, 60, and 75-81 for related information.)

Appendix A
Where Can I Get More Information?

READING MATERIAL

For more information about drinking water, contact the US Environmental Protection Agency in Washington, D.C., (800) 426-4791; 9:00 a.m. – 5:30 p.m. Eastern Time, M-F except on federal holidays; E-mail: hot line— SWDA@epamail.epa. gov; Website: http://www.epa.gov/Ebtpages/ wdrinkingwater.html.

SELECTED TITLES FROM EPA

- *Lead and Your Drinking Water,* EPA/ 810-F-93-001, June 1993
- *Home Water Treatment Units—Filtering Fact from Fiction,* EPA 570/9-90-HHH, September 1990
- *Drinking Water from Household Wells,* EPA 570/9-90-013, September 1990
- *Pesticides in Drinking Water Wells,* EPA 20T-1004, revised September 1990
- *Is Your Drinking Water Safe?* EPA 810-F-94-002, May 1994

- *Water On Tap: A Consumers' Guide to the Nation's Drinking Water.* EPA 815-K-97-002, August 1997
- *Understanding of the Safe Drinking Water Act: Protecting Our Health From Source to Tap,* EPA 810-K-99-004
- *25 Years of the Safe Drinking Water Act: Protecting Our Health From Source to Tap,* EPA 810-K-99-004
- *Lead in School Drinking Water,* EPA 570/9-89-001, January 1989
- *Strengthening the Safety of Our Drinking Water: A Report on Progress and Challenges and an Agenda for Action,* EPA 810/R-95-001, March 1995

OTHER SOURCES

For information on home treatment devices contact:

Water Quality Association
Consumer Affairs Department
(708) 505-0160
Website: http://www.wqa.org
In Canada, write to:
Health and Welfare Canada
Tunney's Pasture
Ottawa, Canada K1A 0L2

Contacting your local water supplier can often provide answers to many of your questions specific to your own system. Its telephone number should be on the water bill.

In the United States, you can also contact your regional US Environmental Protection Agency office:

- *Region 1—Connecticut, Maine, Massachusetts, New Hampshire, Rhode Island, Vermont*
  Environmental Protection Agency
  1 Congress Street
  Suite 1100
  Boston, MA 02114-2023
  Phone: (617) 918-1111

- *Region 2—New Jersey, New York, Puerto Rico, US Virgin Islands*
  Environmental Protection Agency
  290 Broadway
  New York, NY 10007-1866
  Phone: (212) 637-3000

- *Region 3—Delaware, District of Columbia, Maryland, Pennsylvania, Virginia, West Virginia*
  Environmental Protection Agency
  1650 Arch St.
  Philadelphia, PA 19103-2029
  Phone: (215) 814-5000
  Toll Free: (800) 438-2474

- *Region 4—Alabama, Florida, Georgia, Kentucky, Mississippi, North Carolina, South Carolina, Tennessee*
  Environmental Protection Agency
  Atlanta Federal Center
  61 Forsyth Street, SW
  Atlanta, GA 30303-3104
  Phone: (404) 562-9900
  Toll Free: (800) 241-1754

- *Region 5—Illinois, Indiana, Michigan, Minnesota, Ohio, Wisconsin*
  Environmental Protection Agency
  77 W. Jackson Boulevard
  Chicago, IL 60604-3507
  Phone: (312) 353-2000

- *Region 6—Arkansas, Louisiana, New Mexico, Oklahoma, Texas*
  Environmental Protection Agency
  Fountain Place 12th Floor, Suite 1200
  1445 Ross Avenue
  Dallas, TX 75202-2733
  Phone: (214) 665-2200

- *Region 7—Iowa, Kansas, Missouri, Nebraska*
  Environmental Protection Agency
  901 North 5th Street
  Kansas City, KS 66101
  Phone: (913) 551-7003
  Toll Free: (800) 223-0425

- *Region 8—Colorado, Montana, North Dakota, South Dakota, Utah, Wyoming*
  Environmental Protection Agency
  999 18th Street, Suite 500
  Denver, CO 80202-2466
  Phone: (303) 312-6312
  Toll Free: (800) 227-8917

- *Region 9—Arizona, California, Hawaii, Nevada, American Samoa, Guam, Trust Territories of the Pacific*
  Environmental Protection Agency
  75 Hawthorne Street
  San Francisco, CA 94105
  Phone: (415) 744-1305

- *Region 10—Alaska, Idaho, Oregon, Washington*
  Environmental Protection Agency
  1200 Sixth Avenue
  Seattle, WA 98101
  Phone: (206) 553-1200
  Toll Free: (800) 424-4372

Another good source of information is:
American Water Works Association
6666 W. Quincy Avenue
Denver, CO 80235-3098
(303) 794-7711
Website: http://www.awwa.org

Educators, ask for the Public Affairs Department. You may also call:
(800) 366-0107—Small Systems hot line
(800) 926-7337—for the AWWA book on
conservation (Item No. 10063), other publications, or a free Publications Catalog.

Below are listed several other national hot lines and clearinghouses on environmental issues:
- Asbestos Ombudsman Office (USEPA) (800) 368-5888
- Hazardous Waste Information (800) 424-9346

- Hazardous Waste Ombudsman (9 a.m. to 4 p.m. Eastern Time) (800) 262-7937
- Indoor Air Quality Information Clearinghouse (800) 438-4318
- National Lead Information Center hotline (8:30 a.m.–5:00 p.m. Eastern Time) (800) 424-5323
- National Radon hot line (800) 767-7236
- National Response Center hot line (to report spills of oil and other hazardous materials) (800) 424-8802
- National Small Flows Clearinghouse (800) 624-8301
- Pollution Prevention Information Clearinghouse (10:00 a.m.–2:00 p.m. Eastern Time) (202) 260-1023
- USEPA Acid Rain hot line (202) 564-9620
- USEPA Wetlands Information hot line (9:00 a.m.–5:30 p.m. Eastern Time) (800) 832-7828

INFORMATION ON THE NET

AWWA (www.awwa.org)

Water utility links
   http://www.awwa.org/utility.cfm
State drinking water programs
   http://www.awwa.org/statinfo.htm
Drinking Water Week materials
   http://www.awwa.org/dww
Fact Sheets
   http://www.awwa.org/pressroom/
   FACSHTS.htm

Stats on Tap
  http://www.awwa.org/pressroom/
  STATSWP5.htm
Xeriscape Library
  http://www.awwa.org/xeriscape
WaterWiser
  http://www.waterwiser.org
Online Bookstore
  http://www.awwa.org/bookstore/timssnet/
  common/tnt_frontpag.cfm

USEPA Home Page (www.epa.gov)

EPA Information Sources
  http://www.epa.gov/epahome/resource.htm
EPA Regions
  http://www.epa.gov/epahome/locate2.htm
EPA Summary of the SDWA
  http://www.epa.gov/region5/defs/html/
  sdwa.htm
EPA major water projects and programs
  http://www.epa.gov/epahome/
  waterpgram.htm

USEPA Office of Groundwater and Drinking Water Home Page (www.epa.gov/safewater)

FAQ site
  http://www.epa.gov/safewater/faq/faq.html
Drinking Water Publications
  http://www.epa.gov/safewater/Pubs/
  index.html
SDWA Resources
  http://www.epa.gov/safewater/sdwa/
  sdwa.html

Basics of Drinking Water and Health
  http://www.epa.gov/safewater/dwhealth.html
Drinking Water Health Advisories
  http://www.epa.gov/ost/drinking
Links to CCRs
  http://www.epa.gov/safewater/dwinfo.htm

Federal Centers for Disease Control and Prevention (www.cdc.gov)

Crypto information
  http://www.cdc.gov/ncidod/dpd/parasites/
  cryptosporidiosis/default.htm
E. coli information
  http://www.cdc.gov/ncidod/dbmd/
  diseaseinfo/escherichiacoli_g.htm
Giardia information
  http://www.cdc.gov/ncidod/dpd/parasites/
  giardiasis/default.htm
Lead information
  http://www.cdc.gov/nceh/lead/lead.htm
Waterborne disease information
  http://www.cdc.gov/ncidod/dpd/parasites/
  waterborne/default.htm
Safe Food and Water for Travelers
  http://www.cdc.gov/travel/foodwater.htm

Health Canada (www.hc-sc.gc.ca)

Drinking/Recreation Water Publications
and Fact Sheets
  http://www.hc-sc.gc.ca/ehp/ehd/bch/
  water_quality/publications.htm

Chlorination information
http://www.hc-sc.gc.ca/ehp/ehd/catalogue/
general/iyh/chlorina.htm

Chlorinated water and health effects
http://www.hc-sc.gc.ca/ehp/ehd/bch/
water_quality/chlorinated_water.htm

Drinking Water Guidelines
http://www.hc-sc.gc.ca/ehp/ehd/catalogue/
general/iyh/dwguide.htm

Q&A on Bottle Water
http://www.hc-sc.gc.ca/food-aliment/english/
organization/microbial_hazards/faqs_bottle_
water_eng.html

Q&A on Drinking Water Treatment Units
http://www.hc-sc.gc.ca/ehp/ehd/bch/
water_quality/faq_dwtd.htm

Treatment units for disinfection
http://www.hc-sc.gc.ca/ehp/ehd/catalogue/
gieneral/iyh/disinfection_devices.htm

Treatment units for taste and odor
http://www.hc-sc.gc.ca/ehp/ehd/catalogue/
general/iyh/water_treat/taste.htm

Well water quality
http://www.hc-sc.gc.ca/ehp/ehd/catalogue/
general/iyh/well_water.htm

Drinking water away from home
http://www.hc-sc.gc.ca/ehp/ehd/catalogue/
bch_pubs/dw_away.htm

Water Facts and Tips
http://www.hc-sc.gc.ca/ehp/ehd/bch/
water_quality/facts_tips.htm

Environment Canada (www.ec.gc.ca)

Water Home Page
  http://www.ec.gc.ca/water/
includes FAQs, educational materials, and information on the nature of water, water policy and legislation, and water management.
Clean Water Home Page
  http://www.ec.gc.ca/envpriorities/
  cleanwater_e.htm
Water publications
  http://www.ec.gc.ca/water/en/info/pubs/
  e_pubs.htm

Canadian Water and Wastewater Association (http://www.cwwa.ca)

Links to International, Provincial, National and Municipal authorities and organizations
  http://www.cwwa.ca/links.htm

Appendix B
Information About Inorganic Chemicals Found in Drinking Water

This material is adapted from: Sandeen, W. G., "Groundwater Resources of Rusk County, Texas, U.S. Geological Survey, Open File Report 83-757 (1983). *Note: Secondary regulations of the U.S. Environmental Protection Agency mentioned below are nonenforced recommendations.*

ALKALINITY

Description and Sources

Alkalinity is a measure of the capacity of a water to neutralize a strong acid, usually to pH of 4.2. Alkalinity in natural waters usually is caused by the presence of bicarbonate and carbonate ions and to a lesser extent by hydroxide and minor acid radicals such as borates, phosphates, and silicates. Carbonates and bicarbonates are common to most natural waters because of the abundance of carbon dioxide and carbonate minerals in nature. The alkalinity of natural waters varies widely but rarely exceeds 400 to 500 mg/L as $CaCO_3$.

Effect on Drinking Water

Alkaline waters may have a distinctive unpleasant taste.

CALCIUM (Ca)

Description and Sources

Calcium is widely distributed in the common minerals of rocks and soils and is the principal cation in many natural fresh waters, especially those that contact deposits or soils originating from limestone, dolomite, gypsum, and gypsiferous shale. Calcium concentrations in freshwaters usually range from zero to several hundred mg/L. Larger concentrations are not uncommon in waters in arid regions, especially in areas where some of the more soluble rock types are present.

Effect on Drinking Water

Calcium contributes to the total hardness of water. Small concentrations of calcium carbonate combat corrosion of metallic pipes by forming protective coatings. Calcium in domestic water supplies is objectionable because it tends to cause incrustations on cooking utensils and water heaters and increases soap or detergent consumption in waters used for washing, bathing, and laundering.

CHLORIDE (Cl)

Description and Sources

Chloride is relatively scarce in the earth's crust but is the predominant anion in sea water, most petroleum-associated brines, and in many natural freshwaters, particularly those associated with marine shales and evaporites. Chloride salts are very soluble and once in solution tend to stay in solution. Chloride concentrations in natural waters vary from less than 1 mg/L in stream runoff from humid areas to more than 100,000 mg/L in groundwaters and surface waters in arid areas. The discharge of human, animal, or industrial wastes and irrigation return flows may add significant quantities of chloride to surface and groundwaters.

Effect on Drinking Water

Chloride may impart a salty taste to drinking water and may accelerate the corrosion of metals used in water-supply systems. According to the National Secondary Drinking Water Regulations of the US Environmental Protection Agency, the maximum contaminant level of chloride for public water systems is 250 mg/L.

DISSOLVED SOLIDS

Description and Sources

Theoretically, dissolved solids are dry residues of the dissolved substances in water. In reality, the term "dissolved solids" is defined by the method used in the determination. In most waters, the dissolved solids consist predominantly of silica, calcium, magnesium, sodium, potassium, carbonate, bicarbonate, chloride, and sulfate, with minor or trace amounts of other inorganic and organic constituents. In regions of high rainfall and relatively insoluble rocks, waters may contain dissolved-solids concentrations of less than 25 mg/L; but saturated sodium chloride brines in other areas may contain more than 300,000 mg/L.

Effect on Drinking Water

Dissolved-solids values are used widely in evaluating water quality and in comparing waters. The following classifications based on the concentrations of dissolved solids commonly is used by the US Geological Survey.

Classification	Dissolved-solids concentration (mg/L)
Fresh	<1,000
Slightly saline	1,000–3,000
Moderately saline	3,000–10,000
Very saline	10,000–35,000
Brine	<35,000

The National Secondary Drinking Regulations (US Environmental Protection Agency) set a dissolved-solids concentration of 500 mg/L as the maximum contaminant level for public water systems. This level was set primarily on the basis of taste thresholds and potential physiological effects, particularly the laxative effect on unacclimated users. Although drinking waters containing more than 500 mg/L are undesirable, such waters are used in many areas where less mineralized supplies are not available without any obvious ill effects.

FLUORIDE (F)

Description and Sources

Fluoride is a minor constituent of the earth's crust. The calcium fluoride mineral fluorite is a widespread constituent of resistate sediments and igneous rocks, but its solubility in water is negligible. Fluoride commonly is associated with volcanic gases, and volcanic emanations may be important sources of fluoride in some areas. The fluoride concentration in fresh surface waters usually is less than 1 mg/L; but larger concentrations are not uncommon in saline water from oil wells, groundwater from a wide variety of geologic terrain, and water from areas affected by volcanism.

Effect on Drinking Water

Fluoride in drinking water decreases the incidence of tooth decay when the water is consumed during the period of enamel calcification. Excessive quantities in drinking water consumed by children during the period of enamel calcification may cause a characteristic discoloration (mottling) of the teeth. Thus, USEPA has an upper allowable limit for fluoride in drinking water.

HARDNESS

Description and Sources

Hardness of water is attributable to all polyvalent metals but principally to calcium and magnesium ions. Water hardness results naturally from the solution of calcium and magnesium, both of which are widely distributed in common minerals of rocks and soils. Hardness of waters in contact with limestone commonly exceeds 200 mg/L. In waters from gypsiferous formations, a hardness of 1,000 mg/L is not uncommon.

Effect on Drinking Water

Excessive hardness of water for domestic use is objectionable because it causes incrustations on cooking utensils and water heaters and increased soap or detergent consumption.

IRON (Fe)

Description and Sources

Iron is an abundant and widespread constituent of many rocks and soils. Iron concentrations in natural waters are dependent upon several chemical processes including oxidation and reduction; precipitation and solution of hydroxides, carbonates, and sulfides; complex formation especially with organic material; and the metabolism of plants and animals. Dissolved-iron concentrations in oxygenated surface waters seldom are as much as 1 mg/L. Some groundwaters, unoxygenated surface waters such as deep waters of stratified lakes and reservoirs, and acidic waters resulting from discharge of industrial wastes or drainage from mines may contain considerably more iron. Corrosion of iron casings, pumps, and pipes may add iron to water pumped from wells.

Effect on Drinking Water

Iron is an objectionable constituent in water supplies for domestic use because it may adversely affect the taste of water and beverages and stain laundered clothes and plumbing fixtures. According to the USEPA National Secondary Drinking Water Regulations the maximum contamination level of iron for public water systems is 0.3 mg/L.

MAGNESIUM (Mg)

Description and Sources

Magnesium ranks eighth among the elements in order of abundance in the earth's crust and is a common constituent in natural water. Ferromagnesian minerals in igneous rock and magnesium carbonate in carbonate rocks are two of the more important sources of magnesium in natural waters. Magnesium concentrations in freshwaters usually range from zero to several hundred mg/L; but larger concentrations are not uncommon in waters associated with limestone or dolomite.

Effect on Drinking Water

Magnesium contributes to the total hardness of water. Large concentrations of magnesium are objectionable in domestic water supplies because they can exert a cathartic and diuretic action upon unacclimated users and increase soap or detergent consumption in waters used for washing, bathing, and laundering.

MANGANESE (Mn)

Description and Sources

In chemical behavior and occurrence in natural water, manganese resembles iron. Manganese is much less abundant in rocks, however. As a result, the concentration of manganese in water is generally less than that of iron. Under reducing conditions (the lack of oxygen dissolved in water) in water containing dissolved carbon dioxide, manganese dissolves as manganous ion. This sometimes occurs in groundwaters and in water near the bottom of lakes and reservoirs. Manganous ion is more stable in water in the presence of dissolved oxygen than ferrous (reduced) iron under similar circumstances. Manganese concentrations greater than 1 mg/L may result where manganese-bearing minerals are attacked by water under reducing conditions or where some types of bacteria are active.

Effect on Drinking Water

Manganese is an essential trace element for humans. It plays an important role in many enzyme systems. Chronic toxicity has not been reported. With surface waters averaging less that 0.05 mg/L of manganese in several surveys, the potential harm from this source is virtually nonexistent. The main problem with manganese in drinking water has to do with undesirable taste and discoloration (black) of the water. The USEPA secondary drinking water standard for manganese in drinking water is 0.05 mg/L to prevent these problems.

NITROGEN (N)

Description and Sources

A considerable part of the total nitrogen of the earth is present as nitrogen gas in the atmosphere. Small amounts of nitrogen are present in rocks, but the element is concentrated to a greater extent in soils or biological material. Nitrogen is a cyclic element and may occur in water in several forms. The forms of greatest interest in water in order of increasing oxidation state, include organic nitrogen, ammonia nitrogen (NH_3-N), nitrite nitrogen (NO_2-N), and nitrate nitrogen (NO_3-N). These forms of nitrogen in water may be derived naturally from the leaching of rocks, soils, and decaying vegetation; from rainfall; or from biochemical conversion of one form to another. Other important sources of nitrogen in water include effluent from wastewater treatment plants, septic tanks, and cesspools and drainage from barnyards, feedlots, and fertilized fields. Nitrate is the most stable form of nitrogen in an oxidizing environment and is usually the dominant form of nitrogen in natural waters. Significant quantities of reduced nitrogen often are present in some groundwaters, deep unoxygenated waters of stratified lakes, and reservoirs.

Effect on Drinking Water

Nitrate and nitrite are objectional in drinking water because of the potential risk to bottle-fed infants for methemoglobinemia, a sometimes fatal illness related to the impairment of the oxygen-carrying ability of the blood.

pH

Description and Sources

The pH of a solution is a measure of its hydrogen ion activity. By definition, the pH of pure water at a temperature of 25°C is 7.00. Natural waters contain dissolved gasses and minerals, and the pH may deviate significantly from that of pure water. Rainwater not affected significantly by atmospheric pollution generally has a pH of 5.6 because of the solution of carbon dioxide from the atmosphere. The pH range of most natural surface waters and groundwaters is about 6.0 to 8.5. Many natural waters are slightly basic (pH>7.0) because of the prevalence of carbonates and bicarbonates, which tend to increase the pH.

Effect on Drinking Water

The pH of a domestic water supply is significant because it may affect taste, corrosion potential, and water-treatment processes. Acidic waters may have a sour taste and cause corrosion of metals and concrete. The USEPA National Secondary Drinking Water Regulations set a pH range of 6.5 to 8.5 as the maximum contaminant level for public water systems.

POTASSIUM (K)

Description and Sources

Although potassium is only slightly less common than sodium in igneous rocks and is more abundant in sedimentary rocks, the concentration of potassium in most natural waters is much smaller than the concentration of sodium. Potassium is liberated from silicate minerals with greater difficulty than sodium and is more easily adsorbed by clay minerals and reincorporated into solid weathering products. Concentrations of potassium more than 20 mg/L are unusual in natural fresh waters, but much larger concentrations are not uncommon in brines or in water from hot springs.

Effect on Drinking Water

Large concentrations of potassium in drinking water may act as a cathartic, but the range of potassium concentrations in most domestic supplies seldom cause these problems.

SILICA (SiO$_2$)

Description and Sources

Silicon ranks second only to oxygen in abundance in the earth's crust. Contact of natural waters with silica-bearing rocks and soils usually results in a concentration range of about 1 to 30 mg/L; but concentrations as large as 100 mg/L are common in waters in some areas.

Effect on Drinking Water

Silica in some domestic water supplies may inhibit corrosion of iron pipes by forming protective coatings.

SODIUM (Na)

Description and Sources

Sodium is an abundant and widespread constituent of many soils and rocks and is the principal cation in many natural waters associated with argillaceous sediments, marine shales, and evaporites, and in sea water. Sodium salts are very soluble and once in solution tend to stay in solution. Sodium concentrations in natural waters vary from less than 1 mg/L in stream runoff from areas of high rainfall to more than 100,000 mg/L in groundwaters and surface waters associated with halite deposits in arid areas. In addition to natural sources of sodium, wastewater, industrial effluents, oilfield brines, and deicing salts may contribute sodium to surface and groundwaters.

Effect on Drinking Water

Sodium in drinking water may be harmful to persons suffering from cardiac, renal, and circulatory diseases and to women with toxemias of pregnancy. Large sodium concentrations are toxic to most plants.

SPECIFIC CONDUCTANCE

Description and Sources

Specific conductance is a measure of the ability of water to transmit an electrical current and depends on the concentrations of ionized constituents dissolved in the water. Many natural waters in contact only with granite, well-leached soil, or other sparingly soluble material have a low conductance.

Effect on Drinking Water

The specific conductance is an indication of the degree of mineralization of a water and may be used to estimate the concentration of dissolved solids in the water.

SULFATE (SO_4)

Description and Sources

Sulfur is a minor constituent of the earth's crust but is widely distributed as metallic sulfides in igneous and sedimentary rocks. Weathering of metallic sulfides such as pyrite by oxygenated water yields sulfate ions to the water. Sulfate is also dissolved from soils and evaporite sediments containing gypsum or anhydrite. The sulfate concentration in natural fresh waters may range from zero to several thousand mg/L. Drainage from mines may add sulfate to waters by virtue of pyrite oxidation.

Effect on Drinking Water

Sulfate in drinking water may impart a bitter taste and act as a laxative on unacclimated users. According to the USEPA National Secondary Drinking Water Regulations the maximum contaminant level of sulfate for public water systems is 250 mg/L.

ZINC (Zn)

Description and Sources

In general, in streams and rivers, zinc is concentrated in sediments, but concentrations are quite low in running filtered water. In areas of soft, acidic water, however, pickup in the distribution system has been noted when comparing water samples from the treatment plant with samples at consumer's taps. This may result from the water flowing through galvanized iron pipes. In rocks, zinc is most commonly present in the form of the sulfide sphalerite, which is the most important zinc ore. Zinc may replace iron or magnesium in certain minerals and it may be present in carbonate sediments. In the weathering process, soluble compounds of zinc are formed and the presence of at least traces of zinc in water is common.

Effect on Drinking Water

Zinc is considered an essential trace element in human and animal nutrition. The recommended daily dietary allowances for zinc are as follows: adults, 15 milligrams each day, growing children over a year old, 10 milligrams each day, and additional supplements during pregnancy and breast feeding (check with your own doctor). As far as drinking water is concerned, the secondary

USEPA drinking water standard of 5 mg/L, assuming the common intake of two liters of water each day, would result in an intake of 10 milligrams of zinc each day, which is less than the estimated adult dietary requirement for zinc. Concentrations of 40 milligrams in a liter of water gives the water a strong metallic taste (the technical description is **astringent**).

Appendix C
Fascinating Facts
About Water in
General

LIQUID WATER

C1. How heavy is water?

One US gallon (3.8 liters) of water weighs about 8.3 pounds (3.8 kilograms). If you use 100 gallons (380 liters) of water in your home each day (a reasonable amount for two people), you are using more than 830 pounds (450 kilograms) of manufactured product each day, about 150 tons (135 metric tons) each year. A Canadian household of two would use about 180 tons (165 metric tons) of water each year, so on average they would use about the same number of gallons each day as the US couple.

Water is so heavy that on a tonnage production basis the water supply industry is by far the largest in the United States and Canada. In the

United States, it produces in 16 hours as great a tonnage as the output of the oil industry in a year, in a single day as much tonnage as the steel industry does in an entire year, and in a week a tonnage equal to the yearly output of all of the bituminous coal producers—truly a monumental daily quantity of product.

C2.

I have a quarter-acre (0.1-hectare) lot (180 feet by 60 feet [55 meters by 18 meters]). If it rains 1 inch (25 millimeters), how much water falls on my lot?

About 7,000 US gallons (26,500 liters), or nearly 30 tons (27 metric tons) of water. If you had a half-acre (0.2-hectare) lot, it would be twice as much, and so forth.

C3.   Where is the most polluted place on earth?

According to *National Geographic* magazine, the shore of a small lake (Lake Karachay) in Russia's heavily industrialized Ural Mountains is considered to be the most contaminated spot in the world. In the early 1950s, the lake became the dumping area for radioactive wastes from the nearby nuclear weapons production complex at Chelyabinsk. Some have estimated that you could get a lethal dose of radiation by just standing on the shore for an hour.

C4.   Why does water swirl when it goes down the drain?

You might have heard that the swirling is caused by the earth spinning on its axis. The technical name for this is the **Coriolis** effect, named after the nineteenth-century French mathematician who figured it out. This is not correct, however. Although the Coriolis effect on a global scale is a major driving force for ocean currents and weather, it is extremely small on the scale of your bathtub or sink. Furthermore, it would take many hours of water standing in a bowl until it was still enough to see this effect. Even then, the swirling would be quite slow.

The swirling that occurs is caused by the same effect you see when an ice skater is spinning. When the skater pulls her or his arms in close to the body, the skater spins faster. In the water bowl, even though the water looks quiet, there are slight circulating motions, the movement being left over from filling the bowl or tub with water. Because the drain opening is much smaller than the bowl, as the water gets near the small drain, the movement gets faster so you can see it, like the ice skater pulling her or his arms close to the body.

If you want to test this, fill a sink with water and stir it clockwise with your hand. This test works best if you use a large sink that has rounded corners. Then wait two or three minutes until the water looks still, open the drain and watch the water swirl (a drop of food coloring may help you see the swirling). Now do this test again, only this time stir the water counterclockwise. This time the water will swirl the other way as it goes down the drain.

For more information on this subject see: <http://www.ems. psu.edu/~fraser/Bad/ BadCoriolis.html>.

C5. Why is water in the ocean blue?

Sunlight contains all colors of the rainbow, but when you see a color it is because that color of light bounces off the object and reaches your eye, the other colors don't reflect. The red, blue, yellow, purple, and other colors of light in sunlight don't reflect off a green leaf, only green light does.

The sky is blue on a nice day because only blue light reflects off the particles in the atmosphere. (In outer space, the sky is black because there are no particles for the light to bounce off. Remember, black is not a color. You see black because no light reaches your eye). The ocean is blue because as the blue light comes out of the sky, it bounces off the water and into your eye, fooling your eye into thinking the water is blue, when it actually has no color. Notice next time it's really cloudy that the ocean loses most of its color because so little blue light is coming out of the sky.

If the ocean is clear and shallow, the bottom can influence the color, and the water can have different shades of blue, even green sometimes. Swimming pools are often painted blue on the bottom to make the water in them appear blue.

C6. Why is ocean water salty?

Rainwater doesn't contain any salt, but when it falls on the ground, salt from the soil dissolves in the water as it flows back down to the ocean. When this water evaporates from the ocean (see Question 149), the salt stays behind. This has been going on for more than a billion years. Over that very long period of time the ocean got more salt in it with each cycle.

Some salt is removed from the ocean; for example, when water in shallow ocean inlets evaporates and salt beds form. This salt is no longer considered to be part of the ocean salt.

C7. Why does water flow downhill?

Gravity tries to pull everything in toward the center of the earth. When you drop a ball, it moves closer to the center of the earth (we say down but it is really *in* toward the center of the earth) until it can move no farther because it is resting on the ground. When the ground is on an angle, the high part is somewhat farther from the center of the earth than the low part. Gravity pulls water in the direction that will move it closer to the center of the earth, or downhill.

C8. Why should I wash my hands in hot water?

Many of the things you are trying to get off your hands are somewhat greasy, dirt and germs, for example. You know what happens when you put a container of liquid grease in the refrigerator; it gets hard. You want the greasy things on your hands to be soft, so the soap or detergent can get at it and lift if off of your skin. Using hot water helps the cleanser work well and clean your hands better.

C9. Is it true that a microwave oven heats food and beverages by just flipping water molecules back and forth? How is that possible?

A water molecule has one oxygen atom with two hydrogen atoms connected to it, thus, H_2O. The shape of the molecule is like a flat V, with the oxygen at the point and the two hydrogen atoms hanging off it. Oxygen has two negative charges and each hydrogen has a positive charge, so a water molecule can be thought of as a little magnet (although it isn't one), with one end negative and the other end positive. When the microwave is on, the waves effect the way the water molecules line up. Because the microwaves are produced by alternating current (AC), they change direction about 2 billion times each second. Each time they change direction, the microwaves flip the water molecules over to a new direction. Because water molecules are 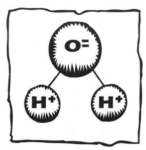 so small, there are trillions upon trillions of them in any food or beverage you put in the microwave. All of them (as well as some other molecules) being flipped back and forth so rapidly by the microwaves causes enough heating to defrost, cook, and heat food and beverages.

C10. Why does it take longer to cook things in boiling water when I am camping in the mountains?

The temperature at which water boils depends on the atmospheric pressure. Atmospheric pressure depends on the weight of the air above any location. At sea level, the weight of the atmosphere "pushes down" with a pressure of 14.7 psi (101.3 KPa) and water boils at 212°F (100°C). Go up into the mountains, however, and you are now above part of the atmosphere, so there is less air above you to make pressure. The lower the pressure, the lower the temperature at which water boils. The temperature at which water boils goes down about 2°F (1°C) for each 1,000 feet (305 meters) of elevation. If you are camping in the mountains at 10,000 feet (3,050 meters), water would boil at only 201°F (94°C). At this lower temperature, it will take longer to cook things.

C11. Are raindrops shaped like teardrops?

According to the US Geological Survey's Internet site, small raindrops (with a diameter of less than 0.1 inch) are shaped like a sphere; larger ones are shaped more like a hamburger bun. When raindrops get larger than a diameter of 0.4 inch, they get distorted into a shape rather like a parachute with a tube of water around the base, and then they break up into smaller drops.

C12.

I heard a joke that if scientists discovered the universal solvent, they couldn't keep it in anything because it would dissolve the container. I also heard that water is the universal solvent. How can that be true? Water is easy to store.

To be exact, if even one molecule of a substance dissolves in water, it is said to be soluble in water. Containers like glass or plastic do dissolve a little, but so little that it doesn't make any difference; they don't leak. Water is considered to be the universal solvent because at least a little of any material will dissolve in it.

C13.

I saw a TV demonstration in which a container was slowly filled with water, and the water rose to a level above the top of the container without spilling. How is this possible?

Water molecules at the surface form a thin "skin" on the top of the water, and this skin is what holds the water above the top of the container. This is called "surface tension." If you break the skin, water will run down the side of the container.

Try this yourself. Fill a container partially with water. When the surface is still, carefully lay a sewing needle on the surface. The skin will hold the needle up and allow it to float on the surface.

FROZEN WATER

C14. I have heard that it is important that ice floats instead of sinks. What difference does this make?

If ice sank when it was formed, each winter the ice would go to the bottom of all the lakes, ponds, and reservoirs. Even icebergs would sink to the bottom of the ocean. When spring came, not enough heat would get to the bottom to melt the ice down there. So after a while, almost all the water on the earth would be frozen. Life as we know it would not have developed on the planet.

C15. How do ice skates work to allow the skater to glide over the ice?

When you squeeze ice hard, it melts. The whole weight of an ice skater is pressing on the thin blade of the skates. This causes so much pressure on the ice, that the ice melts under the blade, making a thin film of

water. The skater glides on this thin film of water. This happens even when the air temperature is very cold.

C16. What is the difference between sleet and hail?

Because it is so cold high up in the air where precipitation starts, all precipitation starts out frozen. If the air near the ground is cold enough, the precipitation never melts to become rain but stays frozen as sleet.

Hail is different. It is created in strong thunderstorms that contain strong updrafts. As the precipitation that will become hail first nears the ground, it is blown back up into the sky where it picks up a bit more moisture that freezes on the outside making the hailstone a bit bigger. Then it starts down again. In very strong storms, the hail goes back up several times, getting bigger each time. When it finally gets so heavy that the updrafts cannot blow it upward anymore, it falls to the ground. If you pick up a large hailstone and cut through it, you will see rings as on a cut tree trunk, each ring indicating another trip up into the sky. Sleet comes when it's cold outside; hail usually comes in the summertime when thunderstorms form. *NOTE: Freezing rain is rain that freezes when it lands on cold ground.*

C17.

Even when the temperature stays below freezing, the depth of the snow on the ground decreases. How is this possible?

When water boils, it changes from a liquid into a gas (steam or water vapor) and the volume of the boiling liquid goes down. Frozen water— snow—can change into a water vapor without melting. The technical name for this process is **sublimation**. Even without it being warm enough to melt the snow, some can change into water vapor, thus the amount of snow gets smaller.

C18.

Why is the ice in glaciers blue?

Sunlight contains all the colors of the rainbow, but we see a color when something only reflects one of the colors, red for a ripe tomato, for example. The ice in glaciers has been under tremendous pressure because of all the weight on top if it, and this has been going on for a long, long time. As time passed, this pressure changed the nature of the tiny ice crystals so that the only kind of light they will reflect is blue. This does not happen when you make ice in the refrigerator. It is not under high pressure, so the ice crystals don't look blue.

C19. Blue ice®, Igloo ice®, and similar products are used to keep things cold. What's in them?

To keep something cold, as it picks up heat and tries to warm up, that heat must be taken away. To take away the unwanted heat, one must have something else cold, something that you don't care if it gets warm. For example, if you have a soft drink with ice cubes in it, you want the soft drink to stay cold, but you don't care if the ice cubes melt. The ice cubes take up the unwanted heat from the soft drink, finally melting. The technical name for whatever takes up the heat is a *heat sink*.

If you add heat to water, its temperature will go up. Frozen water (ice) is different because it can take up a lot of heat before its temperature goes up and it melts. Therefore, it makes a good heat sink. The commercial products you can buy are about 98 percent water. Perhaps coloring, something to stop bacterial growth, or something to prevent leaking are added, but the products are almost all water.

Pure water freezes at 32°F (0°C), but if anything is added to pure water the temperature has to be below 32°F (0°C) for it to freeze. Commercial products frequently add other chemicals to the water in their product to lower the freezing temperature. When you freeze these products and then put them next to food or drinks they can take up a lot of heat, keeping the food and drinks cold.

This idea is not new. Before freezers were common, to make ice cream freeze, you had to get the temperature of the ice cream below 32°F (0°C). To do this, the ice cream maker would be surrounded by water and ice, and salt would be added. The added salt would lower the temperature of the surrounding mixture and draw heat out of the ice cream, so the ice cream would freeze.

C20. What is dry ice?

Water can be a gas (steam or water vapor), a liquid, or a solid (ice). Carbon dioxide also can be a gas, a liquid, or a solid. The difference is that at temperatures when water is a liquid, carbon dioxide is a gas. The way you make carbon dioxide into a solid, known as dry ice, is to cool it down. Water freezes at 32°F (0°C), but carbon dioxide must be much colder to form a solid, -109.3°F (-78.5°C). It is called dry ice because no water is involved. When dry ice warms up, it doesn't form a liquid like water ice does, but it goes straight to a gas by the process called sublimation (See Question C17). Dry ice is often used to make "smoke" or "fog" in staged productions.

C21. I saw a demonstration where someone put a black cloth on a snow bank on a cold, sunny day and the cloth sank down into the snow. Why did that happen?

The sun sends heat and light to the earth as radiation. Many things reflect light into our eye, so we see them. Something black does not reflect any light, so we see it as black (the absence of reflected light). The same thing is true of heat. The black cloth does not reflect radiation and thus keeps both the light and the heat. This causes the black cloth to get warm, so it melts the ice beneath it. A white cloth would not melt the snow because it reflects both the light and heat, so it stays cool.

Some people use a black hose to warm swimming pool water in sunny areas. The black hose lies out in the sun, collects the heat from sunlight, and warms the water being pumped slowly through it.

C22.

On a television science show experiment, one block of ice was looped with a wire, the other with a plastic line. Both the wire and the plastic line had the same size weight hanging down from beneath the blocks of ice. The wire cut through the block of ice but the plastic line didn't. What was going on here?

The pressure under the wire and the plastic line will melt the ice a little, as in Question C15 about ice skates. The melted water then flows up around the wire and plastic line and refreezes behind the wire or plastic line. When water freezes it releases heat. The wire can move (conduct) this heat to its front (lower) side to continue the melting process of the ice. The plastic line cannot conduct heat as well as the wire, so considerably less melting occurs.

C23. Which freezes faster, hot or cold water?

If you live where it gets really cold in the winter and if you put a plastic cup of hot and cold water outside in the cold, the cold water will usually freeze first. On the other hand, in a laboratory, under carefully controlled conditions, the hot water can be made to freeze first. If you have a refrigerator with no freezer, but just a small compartment that holds a couple of ice cube trays, the following may happen. If you fill the ice cube tray with hot water, when you first put it in the little compartment, it will melt the layer of frost under the tray so that the metal bottom of the ice cube tray will touch the metal of the freezing compartment. If you start with cold water in the ice cube tray, this won't happen, and you will have an insulating layer of frost under the tray that will delay freezing. Therefore, under these circumstances, starting with hot water may produce ice cubes faster. However, your tray will most likely freeze to the metal.

C24. Why do ice cubes bulge from the top of the ice cube tray?

Unlike most things that get smaller when they get colder, water gets bigger (expands) by 9 percent when it freezes. Because the ice cube tray has a bottom and four sides that don't move, ice bulges out of the open top when the water gets bigger as it freezes.

Because frozen water (ice) is expanded, it is lighter than water. Therefore, in the winter, ice floats on the surface (as mentioned in Question C14) while the water underneath stays liquid, providing organisms, including fish, with a place to survive during the cold weather.

C25. What makes ice cubes cloudy?

Commercially made ice is stirred as it is being frozen; household ice is not. Without mixing, many more ice crystals form, and air is trapped in the ice. Light rays are distorted by these crystals and air, and this distortion gives home-frozen ice a cloudy appearance.

Appendix D
Complete List
of Questions

CHAPTER 1 HEALTH

GENERAL

1. Is my water safe to drink?
2. What is the definition of safe water? I've heard it called "potable water." How is "potable" pronounced?
3. How can I consider tap water safe after so many got sick in Milwaukee in 1993?
4. What happened in Milwaukee to make so many people sick?
5. From what I hear and read in the media, tap water certainly doesn't sound safe, yet water suppliers say it's okay. Who's right?
6. Is it true that tap water quality is getting worse?
7. If most tap water is safe, why are engineers and scientists still doing so much research and why is the federal

government thinking about more regulations?

8. Can I tell if my drinking water is okay by just looking at it, tasting it, or smelling it?

9. How do I find out if my water is safe to drink?

10. I received a drinking water quality summary from my water supplier. What key information should I look for?

11. My water supplier's annual drinking water quality summary report shows the amounts of some constituents as "not detected" (ND) or "below detection limits" (BDL). Is this the same as zero?

12. My water supplier's annual drinking water quality summary report shows the maximum contaminant level (MCL) for some constituents to be "treatment technique," not a number. What does treatment technique mean?

13. I've received a notice from my water utility telling me that something is wrong. What's that all about? What is a boil water order?

14. I live in an apartment and don't get a water bill. How will I find out if there are any problems with my tap water?

15. Beyond the Milwaukee disaster, does anyone actually get sick from drinking water?

16. Is tap water safer in one area of a community as compared with another?

17. I have read about animals dying after drinking reservoir water. If this can happen, how can I be sure my drinking water is safe?

18. People are allowed to swim and go boating in our reservoir. Should I worry about this?

19. Is tap water suitable for use in a home kidney dialysis machine?

20. Is it true that people who take antacids regularly are more likely to get sick from drinking water?

21. Is the recommended six to eight glasses of water needed each day to maintain good health required to be tap water, or are other drinks okay?

22. I'm moving or traveling to a higher altitude, and I'm concerned about dehydration. How much water should I drink?

23. When I'm working in the yard, I'm tempted to take a drink from my garden hose. Is this safe?

24. Does the author drink tap water?

MICROBIAL CONTAMINANTS

General

25. Is my drinking water completely free of microbes?

26. How are germs that can make me sick kept out of my drinking water?

27. Viruses, bacteria, and protozoan cysts can all make me sick. Which is the hardest to kill?

28. I'm told that I shouldn't drink my well water or that I need to boil it because my water has coliforms in it, but I'm also told that coliforms are harmless. Then I read that food poisoning can occur because of coliforms in meat. What are coliforms and what's going on?

29. How can I kill all the germs in my drinking water?

30. Could my drinking water transmit the AIDS virus?

Cryptosporidium

31. What are *Cryptosporidium* and cryptosporidiosis?

32. Where do the microbes that cause cryptosporidiosis come from?

33. What is the medical treatment for cryptosporidiosis?

34.   What should I do if I think I have cryptosporidiosis?

35.   Are all water systems affected by the threat from *Cryptosporidium*?

36.   Is drinking water the only source of *Cryptosporidium*?

37.   My water supplier found some of the microbes that cause cryptosporidiosis in our water. Will I get sick if I drink water from the tap?

38.   I'm worried about *Cryptosporidium* in my water. Should I have my water tested?

39.   Is my water supplier doing anything to prevent me from getting cryptosporidiosis?

40.   How can I keep from getting cryptosporidiosis?

41.   I am immunocompromised. What should I drink?

42.   Should I drink bottled water to prevent coming down with cryptosporidiosis?

43.   Will a home water treatment device protect me from the microbes that cause cryptosporidiosis?

Travel

44.   If I travel overseas, in which countries is the water safe to drink?

45. When I'm traveling outside of areas where tap water is safe, getting bottled water is inconvenient and I worry about its quality. Is there something I can take with me to purify water?

46. How is water quality protected on airplanes?

47. How is water quality protected on cruise ships?

48. When I travel to a different place in this country, sometimes I have an upset stomach for a couple of days. Is this because something is wrong with the water?

49. I've heard that households in the United States use a lot of water compared with other countries. Is this true?

50. What water does the president of the United States drink at home and when traveling abroad?

51. Is it okay for campers, hikers, and backpackers to drink water from remote streams?

52. What can campers, hikers, and backpackers do to treat stream water to make it safe to drink?

53. If I had no chemicals, no fire, and no filters, is there any way I could make stream or lake water safe to drink?

CHEMICALS

General

54.   Are all chemicals in my drinking water bad for me?

55.   Does drinking water contain calories, fat, sugar, caffeine, or cholesterol?

56.   Does drinking ice water burn fat?

57.   Are chemicals that are found in drinking water naturally (rather than because of pollution) nontoxic?

58.   I read that organic chemicals are bad for my health. What are they, why are they dangerous, and why doesn't my water utility just remove them?

59.   I'm told that our drinking water contains chemicals like cleaning fluid and benzene. What can I do about this while the water company improves treatment?

60.   I've heard that nitrates are bad for infants and pesticides are bad for everyone. How do nitrates and pesticides get into my drinking water?

61.   There is a blue-green stain where my water drips into my sink. What causes this?

62.   Do hazardous wastes contaminate drinking water?

Lead

Fluoride

Chlorine

Aluminum

73.   I hear aluminum is used to treat drinking water. Is this a problem? Does it cause Alzheimer's disease?

74.   Is it safe to cook with aluminum pans?

Radon

75.   What is radon and is it harmful in drinking water?

76.   I'm worried that my drinking water has radon in it and that the radon will get into the air in my home. How can I test the air in my home for radon?

77.   Will a water softener take radon and radium out of my water?

Arsenic

78.   Should I be concerned about arsenic in my drinking water?

79.   If arsenic is in my drinking water, where did it come from and how did it get into the water?

80.   I have heard that arsenic is often found in well water. I have a well. Should I have my water tested for arsenic? If so, where can I get it tested?

81.   My water was tested for arsenic, and it was above the US Environmental Protection Agency's allowable limit. Do treatment systems exist to remove arsenic?

CHAPTER 2 AESTHETICS

TASTE AND ODOR

APPEARANCE

CHAPTER 3 HOME FACTS

GENERAL

91.   Is it okay to use hot water from the tap to make baby formula?

92.   Is it okay to heat water for coffee or tea in a microwave in a styrofoam cup?

93.   Is it okay to drink water that comes out of a dehumidifier?

94.   What should I do to avoid cold-weather problems with my pipes?

95.   How can I locate my home's master valve?

TREATMENT

96.   Should I install home water treatment equipment?

97.   I heard about a water treatment device that uses an electromagnet to treat water. Does this work?

98.   How do I treat my water if my supply fails?

99.   How can I protect my private water supply?

HARD AND SOFT WATER

100.   Is distilled water the "perfect" drinking water?

101.   What is "hard" water?

102.   Should I install a water softener in my home?

BOTTLED WATER

and dancers—to carry bottles of water with them?

114.   Should I buy drinking water from a vending machine?

OTHER USES IN THE HOME

115.   How much water does one person use each day?

116.   Where does the water go when it goes down the drain?

117.   What can I safely pour down the sink or into the toilet?

118.   I have a septic tank. Should I take any special precautions?

119.   Why do hot water heaters fail?

120.   What causes the banging or popping noise that some water heaters, radiators, and pipes make?

121.   How should I fill my fish aquarium?

122.   I have trouble keeping fish alive in my fish pond. Is there anything I can do?

123.   When I try to root a plant or grow flowers from a bulb in my house, the water looks terrible after a while. What will prevent that?

124.   Roses, azaleas, camellias, and rhododendron all require acid conditions. How should I adjust the acid content of my plant water?

125. I live in a very hard water area and I have a water softener. My plants don't seem to like my tap water. What can I do?

COST

126. What is the cost of the water I use in my home?

127. How does the water company know how much water I use in my home?

128. How does the water company know that my water meter is correct?

129. We had a conservation drive in our area and everyone cooperated. Then our water rates went up. Why?

CHAPTER 4 CONSERVATION

130. What activity in my home uses the most water?

131. Some people say I should put a brick in my toilet tank to save water. How does that save water and is it a good idea?

132. I heard that it's a good idea to control the flow of water from my showerhead. How do I measure how fast my shower is using water?

133. Which uses more water, a tub bath or a shower?

134.   I leave the water running while I brush my teeth. Does this waste much water?

135.   I use a lot of water in the kitchen. How can I conserve there?

136.   Why are there aerators on home water faucets?

137.   Many water quality problems in the home—lead, red water, sand in the system, and so forth—are cured by flushing the system. Isn't that a waste of water?

138.   My water faucet drips. Should I bother to fix it?

139.   How should I water my lawn to avoid wasting water?

140.   I have a private well. I don't need to conserve water do I?

141.   Why do we still have a water shortage when it's been raining at my house?

142.   During times of water shortage, shouldn't decorative fountains be turned off?

143.   During times of water shortage, does not serving water to restaurant customers really help?

144.   Questions 148 and 149 imply that the amount of water on the globe isn't changing, so why should I conserve?

CHAPTER 5 SOURCES

GENERAL

154.   Can wastewater be treated to make it into drinking water?

QUALITY

155.   Which is more polluted, groundwater or surface water?

156.   In towns and cities, what is the major cause of pollution of drinking water sources?

157.   Why isn't urban runoff usually treated before being discharged into drinking water sources?

158.   Does acid rain affect water supplies?

159.   I read about the problem of oil spills. Do they pollute drinking water sources?

160.   I live downstream from a nuclear power plant. Should I worry about radioactivity in my drinking water?

161.   How can I help prevent pollution of drinking water sources?

162.   Why does my water sometimes have sand in it?

CHAPTER 6 SUPPLIERS

GENERAL

163.   How many drinking water suppliers are there in the United States and Canada?

TREATMENT

CHAPTER 7 DISTRIBUTION

174. I have seen work crews cleaning water mains and the water they flush out looks terrible. How can the water be safe if the pipes are so dirty?

175. What are asbestos–cement pipes that carry water under the streets? Are they safe?

176. Why are fire hydrants sometimes called fire plugs?

177. We don't use much water for drinking. Why does all the rest need to be treated so extensively? This seems unnecessarily expensive.

178. How does a water company detect a major leak in the distribution piping system?

179. If leaks are such a waste of water, why doesn't the water utility just fix them?

180. Fixing a broken water pipe looks like a dirty job. How is the inside of the pipe cleaned afterward?

181. What are the causes of low water pressure and should low water pressure concern me?

182. What are cross-connections and why are they a problem?

183. Why is some drinking water stored in large tanks high above the ground?

CHAPTER 8 REGULATIONS AND TESTING

FEDERAL REGULATIONS

184. What federal legislation protects the quality of drinking water?

185. Who enforces the federal standards?

186. Is water that meets federal drinking water standards absolutely safe?

187. How do federal regulatory agencies choose the standard for a chemical in drinking water?

188. Most federal standards are written like this: "Selenium—0.05 mg/L." What does "mg/L" mean?

189. Federal, state, provincial, and local governments will spend a lot of money over the next few years to satisfy the requirements of federal laws about drinking water quality. Will any lives be saved?

TESTING

190. How is my water tested and who tests it?

191. Can water systems be excused from monitoring for some contaminants?

192. Can I test my water at home?

193. I have a private well. Who will test my water?

APPENDIX C FASCINATING FACTS ABOUT WATER IN GENERAL

LIQUID WATER

C1.   How heavy is water?

C2.   I have a quarter-acre (0.1-hectare) lot (180 feet by 60 feet [55 meters by 18 meters]). If it rains 1 inch (25 millimeters), how much water falls on my lot?

C3.   Where is the most polluted place on earth?

C4.   Why does water swirl when it goes down the drain?

C5.   Why is water in the ocean blue?

C6.   Why is ocean water salty?

C7.   Why does water flow downhill?

C8.   Why should I wash my hands in hot water?

C9.   Is it true that a microwave oven heats food and beverages by just flipping water molecules back and forth? How is that possible?

C10. Why does it take longer to cook things in boiling water when I am camping in the mountains?

C11. Are raindrops shaped like teardrops?

C12.   I heard a joke that if scientists discovered the universal solvent, they couldn't keep it in anything because it would dissolve the container. I also heard that water is the universal solvent. How can that be true? Water is easy to store.

C13.   I saw a TV demonstration in which a container was slowly filled with water, and the water rose to a level above the top of the container without spilling. How is this possible?

FROZEN WATER

C14.   I have heard that it is important that ice floats instead of sinks. What difference does this make?

C15.   How do ice skates work to allow the skater to glide over the ice?

C16.   What is the difference between sleet and hail?

C17.   Even when the temperature stays below freezing, the depth of the snow on the ground decreases. How is this possible?

C18.   Why is the ice in glaciers blue?

C19.   Blue ice®, Igloo ice®, and similar products are used to keep things cold. What's in them?

C20.   What is dry ice?

C21.   I saw a demonstration where someone put a black cloth on a snow bank on a cold, sunny day and the cloth sank down into the snow. Why did that happen?

C22.   On a television science show experiment, one block of ice was looped with a wire, the other with a plastic line. Both the wire and the plastic line had the same size weight hanging down from beneath the blocks of ice. The wire cut through the block of ice but the plastic line didn't. What was going on here?

C23.   Which freezes faster, hot or cold water?

C24.   Why do ice cubes bulge from the top of the ice cube tray?

C25.   What makes ice cubes cloudy?

Appendix E
Pronunciation
Guide

alum—AL·um

astringent—a·STRIN·gent

bacteriophage—back·TEER·ee·o·fahj (rhymes with garage)

carcinogen—car·SIN·o·jin

catalytic—cat·a·LIT·ic

chloramine—CHLOR·uh·mean

coriolis—kor·e·O·lis

corrosivity—kor·roh·SIV·it·ee

cryptosporidiosis—KRIP·toe·spo·rid·ee·OH·sus

Cryptosporidium—KRIP·toe·spor·RID·ee·um

cyanobacteria—SIGH·an·o·back·TEER· ee·uh

diuretic—die·u·RET·ic

enterococci—EN·tuh·ro·COCK·sigh

Escherichia—es·chuh·RIK·ee·uh

Giardia—jee·R·dee·uh

giardiasis—jee·r·DIE·uh·sis

haloacetic—hay·loh·ah·SEED·ic

methyl-tertiary-butyl-ether—
    METH·uhl–TUR·she·air·ee–BYU·tuhl–EE·thur

microcystin—micro·SIS·tin

oocyst—OH·oh·cist

osmosis—ahs·MO·sis

pathogen—PATH·o·jin

plutonium—plu·TONE·ee·um

potable—POTE·uh·ble (rhymes with floatable)

potassium—po·TAS·ee·um

precipitate—pre·SIP·uh·tate

sublimation—sub·la·MAY·shon

trihalomethane—tri·HALO·meth·ane

turbidity—tur·BID·it·ee

xeriscape—ZER·uh·scape

Index